SpringerBriefs in Modern Perspectives on Disability Research

Series Editors

This book series on disability research is a comprehensive collection of research on disability and related issues. The series is designed to promote interdisciplinary collaboration and exchange, bringing together scholars and practitioners from different fields to share their perspectives and insights. Disability research is an interdisciplinary field that examines the social, cultural, historical, and political dimensions of disability. It encompasses a wide range of topics, including disability rights, accessibility, assistive technologies, healthcare, education, employment, and social welfare. Disability research scholars employ a range of theoretical and methodological approaches to understand the experiences of people with disabilities, as well as the ways in which disability intersects with other social identities such as race, gender, sexuality, and class.

The series seeks to advance knowledge and understanding of disability by publishing rigorous, innovative, and relevant research. It aims to promote disability rights and social justice by highlighting the ways in which people with disabilities are marginalized and discriminated against in society, and advocating for greater social inclusion and accessibility. The series also seeks to inform policy and practice by disseminating research findings that can help to shape policy decisions and contribute to positive social change.

Sheena Mariam Thomas ·
Ramakrishnan Veerabathiran

Disability, Disaster, and Resilience in Oceania

Navigating Vulnerability and Adaptive Capacities

Sheena Mariam Thomas
Human Cytogenetics and Genomics Laboratory
Faculty of Allied Health Sciences
Chettinad Hospital and Research Institute
Chettinad Academy of Research and Education
Kelambakkam, Tamil Nadu, India

Ramakrishnan Veerabathiran
Human Cytogenetics and Genomics Laboratory
Faculty of Allied Health Sciences
Chettinad Hospital and Research Institute
Chettinad Academy of Research and Education
Kelambakkam, Tamil Nadu, India

ISSN 3004-9709 ISSN 3004-9717 (electronic)
SpringerBriefs in Modern Perspectives on Disability Research
ISBN 978-981-95-8534-2 ISBN 978-981-95-8535-9 (eBook)
https://doi.org/10.1007/978-981-95-8535-9

This Springer imprint is published by the registered company Springer Nature Singapore Pte Ltd.
The registered company address is: 152 Beach Road, #21-01/04 Gateway East, Singapore 189721, Singapore

Preface

Disability, Disaster, and Resilience in Oceania: Navigating Vulnerability and Adaptive Capacities has been composed in an era when the Pacific region finds itself at the crossroads of extremely rapid environmental change, major sociopolitical changes, and continuous fights for rights and acknowledgement. In the entire Oceania, disabled persons still suffer the consequences of disasters the most, not because they are inherently weak, but as a result of their being historically sidelined, a combination of socio-political factors, colonial impacts, and the like, and they have been living with all this for a long time. This book is born of the insistence by disabled people in the region that their knowledge, agency, and contributions be taken into account in the resilience discourse and that disability be reframed in disaster governance. The Pacific Islands, spanning thousands of kilometers across the ocean, with diverse ecosystems and Indigenous cultures, provide a unique perspective on the realities of risk and resilience. For these island communities, the impacts of global warming are not ambiguous or far-off; they live with them every day through rising seas, storms, droughts, infrastructure weakening, and resource shortages. However, in the Pacific, one cannot consider environmental exposure the only reason for disasters. Apart from that, it has to be understood in connection with social systems, family ties, histories of colonization, and the very cultural understandings of power, reliance, and mutual duty that are deeply ingrained.

The book covers a wide range of issues and subjects: the critique of colonialism, critical studies of disability, Indigenous knowledge, perspectives on gender, the question of reducing disaster risk, the analysis of policies, and the experiences of people. It examines how colonial maps and Western knowledge altered the concepts of disability and difference in Oceania, thereby reshaping Indigenous people's understandings of existence, land relations, and community. At the same time, it examines the impact of global pacts such as the Convention on the Rights of Persons with Disabilities, the Sustainable Development Goals, and the Sendai Framework on the Pacific's national policies, which, in turn, remain patchy and incoherent. One of the major themes in this book is the significance of Indigenous knowledge, community practices, and relational resilience. To a large extent, Pacific societies are characterized by kinship, reciprocity, and standard care.

The real-life situations of people with disabilities are the main emotional and analytical point of this research. These people's testimonies show that they are not only more exposed to harm but also have incredible power, creativity, and the ability to adapt to their environments. Their stories question the stereotype of people with disabilities as inactive or reliant; instead, they show how people innovate, organize, and take charge in their communities. Persons with disabilities are those who, for example, promote good practices in accessibility in evacuation shelters, communication strategies, and cultural knowledge, to mention just a few, and thereby actively contribute to the region's resilience. The book also examines the relationships among disability and gender, socio-economic inequality, and geographic isolation. For instance, the case of women with disabilities illustrates the cumulated disadvantages of being subjected to both ableism and patriarchy; being literally and metaphorically trapped by the violence, exclusion from decision-making, and inadequate access to crisis services in their case.

Besides, technological transformation is yet another major topic. With Oceania increasingly digitally connected through networks, satellite systems, and assistive technologies, the region has taken giant steps towards a new era in which participation, advocacy, and inclusion in disaster preparedness are the order of the day. Still, digital divides persist, and technology cannot be the solution to systemic exclusion. The book therefore advocated for cautious, ethical, and culturally aware technology integration within the community-led resilience-building process. The work ultimately concludes that the inclusion of particular disabilities should not be viewed as a minor or lateral issue, but rather recognized as a fundamental principle of justice, sustainability, and resilience in societies. The acknowledgement of people with disabilities as holders of rights, bearers of knowledge, and leaders is not only a way to comply with global policy commitments but also a way to boost the adaptive capacities of Pacific communities already dealing with the uncertainty of the climate. The inclusive resilience not only secures the support of the most disadvantaged but also invigorates the social, cultural, and governance systems that uphold the whole region. The book essentially calls for envisioning a Pacific world of non-discrimination, a world in which resilience is not simply the ability to withstand a crisis but rather a collective commitment to justice, dignity, relationality, and the thriving of all people.

We hope that this work will contribute not only to academic discourse but also to practical action, empowering communities, informing policy, and strengthening advocacy for inclusive and culturally grounded resilience across Oceania.

Acknowledgements The authors thank the Chettinad Academy of Research and Education for their constant support and encouragement.

Competing Interests The authors have no competing interests to declare that are relevant to the content of this manuscript.

Consent for publication All authors have read and approved the manuscript.

Chennai, Tamil Nadu
December 2025

Sheena Mariam Thomas
Ramakrishnan Veerabathiran

About This Book

This interdisciplinary volume critically examines the intersection of disability, disaster risk, and resilience within the unique sociocultural and environmental contexts of Oceania. Home to diverse island nations facing increasing threats from climate change, natural disasters, and infrastructure disparities, Oceania presents a vital site for exploring how disabled communities navigate compounded vulnerabilities.

Drawing on case studies, indigenous knowledge systems, and policy analysis, the book foregrounds the lived experiences of people with disabilities in the Pacific Islands during events such as cyclones, volcanic eruptions, and rising sea levels. It highlights both the systemic barriers to inclusion in disaster preparedness and response, as well as the strength and adaptability of disabled communities in rebuilding and leading resilience efforts. By integrating perspectives from disability studies, environmental justice, and decolonial theory, this volume challenges mainstream disaster discourses that often render disability invisible. It advocates for inclusive disaster governance, culturally grounded resilience strategies, and the co-creation of accessible futures rooted in equity and indigenous sovereignty.

Contents

Abbreviations

AI	Artificial Intelligence
ANZ	Australia and New Zealand
APCD	Asia-Pacific Decade of Persons with Disabilities
Bbb	Build Back Better (Sendai Framework term)
CBO	Community-Based Organisation
CDF	Community Disaster Fund
CEDAW	Convention on the Elimination of All Forms of Discrimination Against Women
COP	Conference of the Parties (UNFCCC)
CRPD	Convention on the Rights of Persons with Disabilities
DPO	Disabled Persons' Organisation
DPPA	Disaster Preparedness and Prevention Act
DRR	Disaster Risk Reduction
ENSO	El Niño–Southern Oscillation
ESCAP	UN Economic and Social Commission for Asia and the Pacific
EWS	Early Warning System
FRA	Framework Review Assessment
FRDP	Framework for Resilient Development in the Pacific
GDP	Gross Domestic Product
GIS	Geographic Information System
HRBA	Human Rights–Based Approach
ICCPR	International Covenant on Civil and Political Rights
ICESCR	International Covenant on Economic, Social and Cultural Rights
ICF	International Classification of Functioning, Disability, and Health
ICT	Information and Communications Technology
ILO	International Labour Organization
IP	Indigenous Peoples
IWGD	Informal Working Group on Disability
MDGs	Millennium Development Goals
MICS	Multiple Indicator Cluster Survey
NDC	Nationally Determined Contributions

NDRRMC	National Disaster Risk Reduction and Management Council
NGO	Non-Governmental Organisation
OCHA	Office for the Coordination of Humanitarian Affairs
OPD	Organisations of Persons with Disabilities
PFRPD	Pacific Framework for the Rights of Persons with Disabilities
PICs	Pacific Island Countries
PRSP	Poverty Reduction Strategy Paper
PwD	Person/People with Disabilities
SDGs	Sustainable Development Goals
SFDRR	Sendai Framework for Disaster Risk Reduction
SIDS	Small Island Developing States
SPC	Pacific Community
SPREP	Secretariat of the Pacific Regional Environment Programme
TEK	Traditional Ecological Knowledge
UN	United Nations
UNDESA	United Nations Department of Economic and Social Affairs
UNDP	United Nations Development Programme
UNDRR	United Nations Office for Disaster Risk Reduction
UNFCCC	United Nations Framework Convention on Climate Change
UNICEF	United Nations Children's Fund
UNISDR	United Nations International Strategy for Disaster Reduction
WHO	World Health Organization

Chapter 1
Colonial Cartographies and the Disabled Body

Abstract This chapter is primarily focused on the colonial cartographies in the Pacific that completely changed not only the geographical space but also the Indigenous people's bodies, their identities, and their ways of knowing. By showing mapping as a tool of imperial domination and control, the chapter argues that European cartographic practices turned the Pacific Islands from a network of kinship and ecological knowledge into an empire with fragmented territories marked by racial, spatial, and bodily hierarchies. The colonial maps, through their visual codes, territorial splits, and classificatory systems, generated, on the one hand, categories of normality and disability, which were exercised through acts of dispossession, militarization, missionary intervention, and medical carcerality. The Indigenous bodies were rendered "abnormal," unproductive, or pathological by these cartographies within the colonial frameworks, particularly as the Western biological and racial paradigms overran the Indigenous concepts of balance, reciprocity, and embodied relationality. The chapter also discusses how the mapping of catastrophes, outbreaks, and environmental changes as signs of Indigenous weakness served to reinforce the narratives of incapacity while, at the same time, erasing Indigenous resilience and adaptive knowledge. Through the use of post-colonial theory, critical disability studies, and Pacific Indigenous epistemologies, the chapter unveils how race and disability intermingled under colonial rule, producing the annihilation of marginalized groups that persist. However, it also points out Indigenous counter-mappings, such as the "Blue Pacific" vision, that not only reclaim oceanic connectivity and reassert sovereignty but also challenge Eurocentric notions of normality, embodiment, and care. Later, the discussion leads to the conclusion that disability in Oceania, on the one hand, is a colonial creation that still has modern impacts, and, on the other hand, is a place where resistance can be grounded in Indigenous views of interdependence, communal health, and toughness.

Keywords Colonial cartography · Disability · Pacific · Counter-mapping · Post-colonial theory

S. M. Thomas and R. Veerabathiran, *Disability, Disaster, and Resilience in Oceania*, SpringerBriefs in Modern Perspectives on Disability Research,
https://doi.org/10.1007/978-981-95-8535-9_1

1.1 Introduction

A map is always an act of power and authority; it is a reflection of superiority and a part of a vision. The Pacific, where the indigenous people claimed land through navigation and kinship, was turned into imaginary imperial domains through colonial maps (Cole & Hart, 2021). Margaret Jolly (2007) states that maps of Oceania were "partial renderings" rather than neutral portrayals, and that these maps were largely determined by the creators, their viewing instruments, and the empire's interests that led them to produce them. The intransigent spatial abstractions of European mapping and indigenous navigational skills, represented by Tupaia of Ra 'iatea, show a notable epistemic gap: While the former was in a way viewing the world from an indifferent and all-encompassing height, the latter was through a relationship, through the body, and through the universe (Jolly, 2007). With 20,000–30,000 islands and islets, Oceania is an area of the world with a total population of around 37 million people spread over 14 nations (Cummins & Capic, 2017).

Cartography matured into an instrument for distinguishing human beings and places in colonial geopolitics. The Enlightenment voyages of Cook and Forster not only opened up the world but also linked geography to the hierarchy of civilizations by classifying races and cultures and mapping coastlines. By the time the nineteenth century rolled around, racialized fantasies were firmly established in geographic space, and difference was made to appear a natural phenomenon through ethnological typologies such as Dumont d'Urville's division of the Pacific Islands into Polynesia, Melanesia, and Micronesia. These maps served two purposes at once: on the one hand, they delineated imperial domains and controlled perceptions of the Pacific; on the other hand, they established racial, moral, and even physical hierarchies that permeated colonial medicine and governance (Jolly, 2007).

The way that colonization and disability intersect to shape the lives of Indigenous peoples is being more recognized on a worldwide scale. According to recent estimates, there are around 370 million Indigenous peoples worldwide (UNDESA, 2009). However, there is a dearth of trustworthy information about the global incidence of disability in Indigenous societies. By emphasizing how Indigenous cultural identity and the lived experience of disability interact, the UN has contributed to the development of the public policy environment. UN Rapporteurs on the Rights of Indigenous Peoples and Persons with Impairments merged their work in 2016 to investigate how impairments affect Indigenous peoples' lives, particularly their continued capacity to participate in and carry out customs, languages, cultures, and traditions (Soldatic & Gilroy, 2018).

In Oceania, disability became a social and political construct that supported those in power, and not just a medical condition, while colonial control lasted. The colonial powers and missionary regimes imposed Western biological frameworks that defined the "normal" body in terms of productivity, reason, and moral order, traits aligned with the so-called European civilization. The colonial structure labeled the Aboriginal, the mentally ill, and the physically or sensorially impaired as "abnormal," unfit, or morally tainted because they were not in line with these traits, at least, the Western ones. Hospitals, asylums, and jail facilities that were, in one way or another, carceral made this distinction between the various classes of bodies. Juliann Anesi (2022) and other scholars point out that these facilities punished, albeit gently, both the Indigenous and disabled people under the guise of civilization and care. The colonial authorities, having installed these systems, totally uprooted the Indigenous people's ideas about spiritual balance, communal care, and bodily presence, thereby introducing the confusing concept of physical difference to be viewed as racial inferiority. Therefore, disability in Oceania has to be considered as a byproduct of colonization and as a process that applied racial, capacity, and civilizational hierarchies onto the body, through which the still existing oppositional terms of able/disabled and normal/abnormal were formed that keep exerting their influence on the views of government and society in the post-colonial period (Anesi, 2022).

From the point of view of the colonialists and cartographers, the Pacific was nothing more than a disconnected "sea of islands"; nonetheless, it was an area thoroughly connected by family ties, genealogies, and fish movements. The pre-colonial Pacific civilizations had very complex and likewise comprehensive views on human beings, health, and potential that were rooted in ecological balance, community welfare, and spirituality. However, these cognitive structures were disrupted by the colonial encounter, which introduced European medical discourses that isolated the person from society, health from the environment, and the body from the spirit. The missionaries and colonial rulers who were mapping the Pacific's peoples not only imposed their own paradigms but also created the racial, gender, and bodily normality hierarchies prevalent in their time (Chang, 2016). Thus, the Pacific gradually shifted from a relational navigation area to one of territorial and physical domination through these external mappings. This relationship was characterized by how race and disability were intertwined, demonstrating colonized Indigenous bodies in a diminished space of inferiority, unproductivity, or needing some punishment. To understand this history, it is essential to consider the epistemic violence colonial mapping enacted, as it changed not only the geographical landscape of Oceania but also indigeneity's ontologies of embodiment and care.

A decolonial and intersectional approach is utilized in this chapter, which later brings together Pacific Indigenous epistemologies, post-colonial theory, and disability studies. Initially, the chapter examines the geopolitics of civil mapping in Oceania to illustrate how the cartographic imagination produced both geographical and physiological inequalities. Following the line of connections between labor, race, and bodily control under imperial dominion, the construction of disability as a colonial category is then analyzed. The preceding section delineates how colonial medicine and missionary institutions facilitated the racializing and medicalizing of difference, and the subsequent section examines their persistence in the post-colonial state and health systems. The study draws from missionary archives, historical records, and current scholarship, including Indigenous standpoints from the area, to highlight how bodies were categorized, resisted, and disciplined as a part of the colonial effort.

The chapter concludes with an analysis of Indigenous counter-mappings and epistemic revival to underscore the aforementioned Pacific philosophies of care, family, and resilience, which are in direct opposition to Eurocentric notions of normality and thereby claim disability as a place of strength and connection. The chapter employs this multidisciplinary framework to suggest that the status of disability in Oceania is an embodied form of resistance against the continuous colonial power maps, rather than a mere absence.

1.2 Colonial Cartographies: Mapping, Power, and Control

Colonial mapping in the Pacific region was not merely a representation; instead, it was a powerful technology that created, controlled, and transformed space in the imperial interest. Colonial cartography in the Pacific region served as a means for Europeans to assert their power, thereby changing the entire oceanic expanse into territories easy to govern and exploit. The maps drawn by cartographers in Britain, France, Germany, and later the United States already divided the Pacific into regions such as Polynesia, Micronesia, and Melanesia, thereby imposing artificial frontiers that completely disregarded Indigenous peoples' navigation methods and spatial relationships. The process of mapping began in the 1800s, with explorers and colonial officials using maps to take over islands, thereby effectively cutting off existing Islander connections across the waters (Matsuda, 2006; Sharma, 2021). The mapping practices of European empires in Oceania not only helped them identify coastlines and trade routes but also defined populations, classified bodies, and justified a new social order. Thus, mapping was the means by which the colonial administration exerted its influence over territory, people's movements, knowledge, and even the criterion for who was recognized as human. The impact was significant for Indigenous peoples, whose physical existence, self-identities, and locations were forced to fit the colonial-imperial viewpoint under "settler" colonial rule (Hattori & Samson, 2023; Woll, 2023).

The power relations characteristic of the maps denoted a "cartographic gaze," viewing the Pacific lands as mere voids to be occupied and ruled, thereby encouraging the extraction of natural resources and the establishment of military bases. For example, at the end of the 1800s, nearly all Pacific Islands were colonized, and later their borders were used by newly independent countries, resulting in a situation in which political and economic interests overshadowed local powers. Indigenous names and knowledge were not included in the maps, thereby making the Islanders invisible and at the same time, indicating the ports and resources that were of strategic importance, hence, a visual control was practiced (Franklin, 2006; Pickles, 2012).

The vision went on through the twentieth century, which came through the Cold War mappings when the United States and allies considered Oceania a strategic denial zone against communism, and so they were conducting nuclear tests on islands like Bikini and Moruroa. After gaining their independence, the Pacific Islands have been classified as "Asia–Pacific" and "Indo-Pacific," which eventually led to their being viewed as minorities, framed within the economies of coastal areas. The exclusions in colonial maps, such as the cultural values of reefs and fisheries, which deny the

Table 1.1 Colonial mapping practices and their effects on indigenous geographies

Colonial mapping practice	Description	Effect on indigenous peoples
Imposition of European spatial abstractions	Maps were "partial renderings," shaped by empire's interests	Replaced Indigenous navigational knowledge and kinship ties
Division of Pacific into Polynesia, Melanesia, Micronesia	Dumont d'Urville's racial typology	Created racial hierarchies tied to geography
Mapping as imperial domination	Maps served "to assert power… and transform space"	Territories were fragmented for easier control
Cold War strategic mapping	Pacific treated as "strategic denial zone"	Enabled nuclear testing and militarization
Creation of borders (e.g., International Date Line)	Arbitrary temporal and spatial divisions	Disrupted Indigenous temporalities and cultural unity

economic need for outsiders, have been criticized as a kind of power projection that is not manifest but de facto in the geopolitical situation (Chand, 2010; Levering, 2016).

Control over cartography also changed how people perceived time. This is evident in the arbitrarily placed International Date Line, which splits the culturally connected islands of Samoa and the Cook Islands by up to 23 h. Such impositions cut off indigenous peoples whose temporalities were based on oceanic cycles, demonstrating the maps' power to impose not only spatial but also existential dominance. The Pacific islands' counter-mapping, such as the "Blue Pacific" narrative, which reclaims ocean connectivity and stewardship against outsiders' framing, has been a form of resistance (Ganguly, 2021; Kabutaulaka, 2021; Martinsson-Wallin & Thomas, 2014; Trask, 1990). The different colonial mapping practices and their effects on indigenous geographies are illustrated in Table 1.1.

1.2.1 Imperial Knowledge Systems and the Classification of Bodies

The imperial knowledge systems raced and classified the Pacific bodies based on the colonial taxonomies that supported the colonial rule, and the Islanders were divided into Melanesians ("dark islands"), Polynesians ("many islands"), and Micronesians ("little islands") according to the skin tones and physical features perceived to be. These categories, named by French explorer Dumont d'Urville in 1832, were not significantly different from the African colonial ones—the dark-skinned Melanesians with curls were treated as "Black" proxies, while the lighter-skinned Polynesians were considered almost "white." This pseudo-scientific ordering played a decisive role in turning bodies into objects of study and in establishing hierarchies that made the subjugation of the people seem natural (Castellino & Keane, 2009).

The classification did not only stop at race but went further to characteristics, where the colonial gaze pathologized the Islander towards attributing those physiques as primitive or unproductive, while, in turn, the white settlers had their ideal qualities constantly projected. In the case of Guam, the Chamorro people were thought of as "lazy" and trained to be a part of the group; thus, they were not suitable for individual labor, which made it easier for them to be pushed to the bottom of the job ladder, while Filipinos were forced to do the unskilled work under threat of deportation. Such systems made whiteness a measure of productivity as white Americans were the ones who took the skilled positions in the military labor forces, for example, right after World War II (Matsuda, 2006; Trask, 1990).

The whole process of knowledge production was heavily dependent on policy, as academics provided reasons for imperial expansion, blending European standards with local traits to legitimize control. The twentieth-century genetic research and bio-sampling took the Indigenous people apart and preserved their "vanishing" DNA while the nuclear colonialism was going on; on the other hand, this process brought about resistance, for instance, the Havasupai lawsuit over the use of the samples without consent. These imperial epistemes persisted, shaping IR discourses that rendered non-Western voices less audible (Gani & Marshall, 2022; Radcliffe, 2018).

1.2.2 Mapping as Domination: Producing Colonial Space

By dividing the Pacific into smaller, manageable pieces, the colonial maps made the region's potential and resources known to the empire. Woll (2023) asserts that the imperial maps treated entire vast archipelagic worlds as simplified, bordered spaces that permitted and facilitated the drawing of Western geopolitical lines between the powers. Boundary drawing—between the islands, countries, "native districts," and administration zones—was an act of spatial violence that not only denied but also replaced the fluidity of the Indigenous peoples' movements with rigid spatial orders.

Dominating the original mapping by the colonizers, island worlds were depicted as places for military and geological operations, or even as the ground for testing pseudo-scientific racial theories. Micronesian stick charts, a method that expressed Indigenous people's passionate attachment to and intricate knowledge of their environment, were called primitive, while the next step was the imposition of colonial grids, true to imperial hierarchies. According to critical cartographers, every map discloses the politics of visibility: what gets shown, what gets erased, and who is allowed to represent the area (Harley, 1989; Lewis & Wigen, 1997).

1.2.3 Territoriality and the Colonial Imaginary of the Pacific Islands

The colonial imaginary created the Pacific Islands as strange, remote, and stunning places, which in turn encouraged territoriality to the point that it was seen as a cross of justification among the empires. European powers considered the sea as a barrier and divided the islands into protectorates and mandates, like the 1931 Keesings' map that indicated imperial overlays notwithstanding the presence of Indigenous governance in places like Tonga. This imaginary also tore apart interconnected Islander networks; it thus replaced them with fixed borders on a "boundless" sea (Chand, 2010; Prabhakar, 202).

Territorial claims were mainly driven by security, with Britain taking control of the islands to protect the neighboring colonies, while the U.S. "strategic denial" mapped out Micronesia after WWII for the establishment of bases such as Kwajalein. France's military presence in the form of nuclear tests was allowed by its retention of territories in Polynesia; thus, militarism was inscribed into landscapes. The 1990s labeling of the region as an "arc of instability" further territorialized it as Australia's backyard; consequently, it was the region that determined the dominating countries' aid and interventions (Kabutaulaka, 2021; Venkataramanan, 2024).

In the present day, the geographies of the past are still reflected in the strategies of the Indo-Pacific and the Maritime Silk Road, which draw the Pacific into the competition among the great powers without the intervention of the Islander peoples. The Blue Pacific initiative counters this by envisioning a connected "Blue Continent," where the sharing of ocean resources is preferred to the division of ocean territories (Kabutaulaka, 2021).

1.3 Colonialism Among the Disabled Community

Disability theorists have been discussing this all along—by the way, they have even suggested that, for the most part, (post)colonial theorists have used disability as a metaphor while for disability theorists the concept of colonialism has become a pivotal metaphor to render the realities of such abused, oppressed, marginalized, and excluded groups as people with disabilities (Barker & Murray, 2010; Sherry, 2007). The occurrence of such conflations in the discourse of both fields has led to a denial of the "necessary recognition of an uneven biopolitical incorporation" (McRuer, 2010) in their characterization of the inter-relation among bodies and minds through space, history, time, and politics that often govern them in various ways (Soldatic & Grech, 2014).

These negative consequences of the processes of dialogical praxis have been instrumental in the formation of new approaches that are intended to open up and creatively challenge the development of the two fields, keeping them aligned and, even more so, together. Whereas post-colonial theory and the disciplines it has interacted with (e.g., critical theory and cultural studies) have been concerned with race, gender, and ethnicity, the realm of disability has been nearly completely ignored—paradoxical suffocation by the historical subjectification of the non-disabled or, rather, disempowering colonialism and the post-colonial landscape. In post-colonial thinking, references to disability remain highly restricted to the implications of political processes that lead, among other things, to the oppression, marginalization, or denial of the rights of certain groups based on the conditionality of colonization, empire, and imperialism (Sherry, 2007).

Disability is occasionally invoked in post-colonial critiques, but, ironically, there is almost no reference to disabiling processes for disabled people in this area of research. The situation is even worse for the more radical schools of thought connected with post-colonial feminists (Grech & Soldatic, 2017) who have not only neglected the topic of disability altogether in their gender race, ethnicity, colonial histories or intersectionalities discussions, Erevelles and Minear (2010) have declared that what we see are "disability-free" colonial and post-colonial areas (epistemologically and ontologically) which, one could say, confines post-colonial theory and restricts its analytical range. The absence of disability from (post)colonial writings is more than curious, as disability has always been and still is one of the most remarkable and ubiquitous conditions of human existence, cutting across time, space, and politics, while also blurring the boundaries between the discursive and the material (Grech & Soldatic, 2017). More fundamentally, the invisibility of disability limits the potential for multidisciplinary dialogue with fields such as disability studies and the sociology of the body, which could greatly enrich the development of post-colonial theory.

Unfortunately, the hegemonic global North disability studies have not played a role in positively changing it. It is still isolated from the global South and its histories, contexts, and cultures, and the epistemologies of these specific geopolitical spaces, including how disability is ontologically constructed and lived through the entire history of power and empire symbolically framed by the global (Grech, 2011; Meekosha, 2011; Soldatic & Biyanwila, 2006). The latter is the case with the domination of global North epistemologies; it has seldom even touched the "colonial" concept outside the metaphoric terrain of the descriptive. The term "colonialism" is so much associated with its analogical aspects that it cannot appropriately characterize the living condition of disabilities as one of being overpowered and oppressed, such as the case of the medical profession and/or research, and not actively exposing disabilities as a critical location of colonial administrative power, a living experience under colonial rule, or a category of difference created to maintain colonial legitimacy and control (overtly or covertly). Consequently, disability is constantly excluded from this wider contextual history. When disability is put in a historical framework, it is too often and too easily dehistoricized and divorced from these larger colonial connections and continuities, such as the exportation/imposition of disability discourse, epistemologies (e.g. the social model of disability), and practices from the global North to global South in post/neocolonial periods, a theme extensively covered in post-colonial studies, but one that disability studies often remains moved away from. The study by Grech and Soldatic, (2017) highlights the necessity of post-colonialism to disability studies and declares that "post-colonialism can dismantle the all-encompassing traits of Western discourse and thus open up the avenues of normativity questioning and the adoption of responsibility for changes from the domain of knowledge's situatedness" (Grech & Soldatic, 2017).

On the contrary, this does not mean there have not been previous attempts to cross the disability/post-colonial divide. Indeed, a growing number of critical disability scholars have already begun to illustrate the importance of disability as an ideological, epistemological, representational, and experiential (post)colonial experience that has been lived through and with post-colonial anxieties, tensions, discourses, and materialities (see Barker & Murray, 2010; Erevelles, 2011; Parekh, 2007).

1.4 Intersectionality of Race and Disability

Investigating the intersection between race and disability has the potential not only to expand but also to enrich the set of research perspectives (i.e., stakeholder expansion) deemed meaningful. This also leads to thinking about how one would measure and weigh inclusive first-person perspectives against the community's burden, and whether, instead of working with the community, other stakeholders should be considered. There are critical considerations regarding the community's engagement and reciprocity that must be addressed (Gonzales, 2019), including avoiding the imposition of technocentric biases and destabilizing assumptions before intervening, as Bennett et al. (2021) point out. Clarification of the terms race and disability is necessary for researcher-community interaction, as their meanings vary across cultural, communal, and social contexts. Race and disability are identity dimensions that determine technological applications differently. All our case studies define race and disability, but they differ in the level of detail they provide. It would be reasonable to think this might be due to factors such as participant privacy or disclosure. Finally, the ongoing recognition and questioning of power relations cannot be overemphasized. These are revealed among the authors, in the different stakeholder and participant groups, between the technologies and the communities, and through access and cultural practices.

Due to insufficient data, research, and reporting on the intersection of disability and technology, there is a need for greater credibility and trust in situated knowledge (Brinkman et al., 2023; Harrington et al., 2023; Rios et al., 2016; Sabariego et al., 2022). This is to a great extent due to the very nature of knowledge production in the area of disability, its dynamic characteristics, the epistemic bases that underpin it, and the inevitable translation of such information into formal knowledge (Nijs & Heylighen, 2015). What is conventionally regarded as empirical research and learning tends to marginalize and exclude the bulk of the valuable insights and meanings that disability groups create from their everyday experiences. The ASSETS community, by Hofmann et al. (2020), has raised this concern, pointing to three main considerations: ableism, the simplification of disability, and the presence of human interactions and relationships around disability.

Naturally, the studies recognize that race need not be the primary concern of every accessibility project, nor vice versa. Moreover, identity is often a fluid state, and a group of individuals may not be homogeneous in terms of their identities. For instance, it would be wrong to think that all older people are disabled, as it is equally wrong to consider all disabled people as being white, or that different races have the same misunderstanding of disability. For instance, the transition from additive thinking to the use of intersectional concepts in data gathering processes may be a challenge for some within the accessibility hemisphere (Windsong, 2018). The analysis by Harrington et al. (2023) has yielded certain guiding principles that we see as the means to help establish and maintain this research area as one that seeks to amplify the experiences of individuals who are at the intersection of disability and race, rather than a problem–solution orientation:

1. Looking Beyond Academia. The study conducted by Harrington et al. (2023) highlighted that race and disability discussions are no longer confined to the ivory tower of academia but are now taking place mostly outside. Both the disability advocates and the racial justice activists have used different forms of media, blogs, and art to expose the specific problems of racially minoritized individuals with disability. It is a matter of practicing citational justice by recognizing the types of knowledge production that are considered valid and valued as a form of action. The researchers should note the activist-led projects, integrate these works into their syllabi, and invite the activists to give course lectures. Teaching and project collaboration are the domains in which the academy can be transcended to augment our understanding of this intersection.
2. Reduce Assumptions of Participant Defaults. Various meta-reviews have demonstrated that only a small number of HCI accessibility-related papers even mention the concept of race, let alone its interaction with technology (Harrington et al., 2022). We must, as a field, actively move away from the default situation in which the reader is left to presume that the sample is white (or able-bodied) unless explicitly stated. Such defaults limit the inclusiveness of the research process. Therefore, as a discipline, we need to be up to date on the issues of race and disability, since even if a particular study (or body of work) does not include race, it is still necessary to acknowledge when intersectional matters arise in a study, or they will either be cut off or treated as minor footnotes. Researchers must destroy a good deal of the existing defaults and definitions that overlook the subtlety and intricacies of the experiences of those who are situated at the crossroads of racial identity and disability. We prompt more race reporting as a demographic factor and the interaction with it as a construct in terms of research design and analysis of findings, where it is deemed suitable.
3. Overstepping the Limits. Although data reporting (and gathering) is very important, we should still be careful not to engage in mere categorization for its own sake, but rather to address race and disability issues in a very significant manner. To what extent has the race of the survey participants affected their viewpoints,

or what has been the impact of their social positioning on their perspectives? Extending this further leads to the idea of identity's multiple dimensions being only additive (Rankin & Thomas, 2019; Windsong, 2018). Researchers are to refrain from skimming over the social aspects of race and instead be open to what this might reveal concerning the phenomena of sociotechnical systems and technology demands. Treating race simply as a synonym for marginalization or being underserved is an injustice to how racial identity might be involved. Consequently, it also creates an unfair negative image of certain groups in our field of research. When we refer to people with low socioeconomic status, we often think of Black/Brown people, which is a misconception that overlooks the viewpoint of White people in this category. Moreover, the assumption that the poor equals Black or Brown is not only racist in its very nature but also distorted because of the areas we are operating in.

Considerations Going Forward: In the current models, community engagement is necessary to continue (Andrews et al., 2019). This is important, but it is of utmost importance that the engagement with the community's representation continues. A good number of us have been involved in various ways in discussions of social justice within the research community (Leal et al., 2021). The engagements through interactive dialogues across races, disabilities, and parallels become the prevailing issues in the norms when community participation is allowed to thrive in research groups, through volunteer work, or across the whole research activities. The separate areas of research that might not be directly linked to the single project, such as community building, dissemination, teaching, advocacy, and activism, could help dismantle the structures of power and oppression and leave a positive impact on communities. The same case goes for disability work, where people without the input of individuals having the first-hand experience of disability can be problematic. The team's diverse perspectives and experiences in disability, race, and their intersection have been a great asset. Minding the community's needs and making space for people who can speak for it directly, i.e., the disabled persons, the racially minoritized individuals, and the ones at the intersection, are the guiding principles of this work. It is not a coincidence that some of the top researchers who have started to penetrate these areas are the ones who have gone through it and therefore cannot turn a blind eye to the fact that these are the pervasive and significant experiences that our current methods keep silent and ignore (Harrington et al., 2023).

1.5 Cartographies of Catastrophe: Disability, Disaster, and Empire in the Pacific

The imperialists illustrated the misfortunes in the Pacific with their maps to justify their takeover, not only portraying the natives as incapable but also hiding their means of survival through nature's forces, epidemics, and famine. The maps led to the establishment of a condition of weakness by opting for the extraction infrastructures that were not sustainable over the eco-friendly ones, thus linking disability with the racialized notions of catastrophe. Climatic factors such as rising sea levels and storms were among the legacies that had the greatest impact, and non-disabled Islanders were the most affected during the current crises compared with non-disabled Islanders (Fakhruddin et al., 2015; Jupiter et al., 2014).

The colonial infrastructures in Samoa and the other Pacific islands made the situation even worse during the cyclones and floods. This was so because they were building coastal roads, hydropower plants, and water systems that were ill-suited to the local climates. During Cyclone Evan in 2012, the land transport alone suffered damage of USD 25.6 million. The reason for this damage was that bridges and roads were poorly designed and poorly maintained, leaving them highly exposed to storm surges and erosion. Water treatment facilities such as Alaoa suffered losses of USD 2.3 million; their pipes were made of lower-quality material and were located above ground, following the rivers' alignment, which is prone to flooding. The power sector was the most affected—ranked as the most vulnerable—due to the dying diesel stations and damaged hydropower facilities. This was the case with Tanugamanono and the other power plants that had lost capacity due to eroded catchments resulting from land clearing.

There was a sewage spill during floods, contaminating water supplies in Apia. These infrastructures, which were very hard, were built on the basis of stationary climate assumptions and did not take into account the rising sea levels and extreme rainfall that climate change was creating, which is why the communities were disrupted. Mining for construction encroached on the beaches and reefs, thus increasing the coastal vulnerability of low-lying islands like Tuvalu. The Colonial legacies continued to be a part of post-disaster failures in the case of "build-back-better" without vulnerability assessments, thus trapping transportation and utilities in damage cycles. Stakeholder consultations indicated that transport was the most vulnerable due to its very high exposure and the associated repair costs (Chape, 2006; Fakhruddin et al., 2015).

The Pacific islands were gradually emptied by epidemics beginning in the eighteenth century, with measles, influenza, and dysentery among the diseases that wiped out up to 50% of the populations in Tahiti and the Marquesas as soon as Europeans arrived. Besides the existing colonies and their labor demand for coffee picking, the introduction of plantations and their consequent monoculture practices led to the extinction of native fish and to famine through reduced soil fertility and disrupted fishing. The politics of aid helped to portray the Islanders as totally unable to help themselves, and thus, the importing of Indian and Chinese workers took place while the local recovery methods were totally ignored. Massive movement of laborers known as blackbirding in Fiji and Queensland, accompanied by diseases, resulted in the death of such large numbers of people that the colonial governments only took the help of contract labor and did not intend to invest in health infrastructure. The Food Insecurity resulting from the use of imported processed foods eventually led to the NCD epidemic in the modern era; moreover, this has been coupled with announcements by Heads of state regarding crises linked to the economic losses they suffer. The regional policies prioritized levying taxes on unhealthy imports but struggled to implement them across the sectors involved. The famine responses were a means of reinforcing the superiority narratives, as the quarantines segregated "diseased" bodies, which made the Indigenous diets appear unhealthy. The current aid situation is similar, as fiscal measures such as sugar taxes are becoming more widely accepted but still face opposition from the trade agenda (Dodd et al., 2020; Rallu, 2011; Shanks, 2016).

Colonial documents recognized and recorded sacred sites and calamities for the aim of extraction, thus carefully controlling oral knowledge by placing it in a grid format that could not capture the spiritual values of the areas. The priorities of maps were mines rather than graveyards, making the natives' resistance invisible during famines and disease outbreaks. Research on the material left behind by the natives shows that large communities existed before the arrival of Europeans, contradicting the story of depopulation entirely through disease. The pain was represented as racial inferiority, with storms like Evan being recorded for financial losses, while at the same time, people's coping strategies were not. The erasure continued in the vulnerability ratings, which did not take into account soft infrastructure, such as early warnings rooted in Indigenous knowledge, that were present in the area. The natives' power to resist has been shown through their attempts at counter-mapping, which focus on reclaiming holistic relationships with the land. The inequalities in health today go back to such records, and thus, call for or support structural analyses rather than cultural blame (Buscher & Pearce, 2024; Fakhruddin et al., 2015; Rallu, 2011; Short, 2009).

1.6 Conclusion

The colonial maps of the Pacific did not merely delineate coastlines or demarcate territories. They fundamentally changed the way people, bodies, and landscapes are understood. Colonial mapping practices superimposed abstract spatial grids onto relational oceanic worlds and created new hierarchies of race, ability, and civilization that got ingrained in the spheres of medicine, governance, labor, and social relations. The disabled body that was constructed according to the Western biologized norms of productivity, rationality, and moral worth came to be a central figure in the colonial project. The need for control, intervention, and "civilizing" missions was justified through the figure of the disabled body. Colonial powers, through maps, censuses, medical surveys, and territorial partitions, redefined Indigenous bodies and communities as inferior, weak, or degenerate, thus legitimizing the establishment of the systems of surveillance, segregation, and carceral care. Legacies like these were not wiped out with decolonization. The post-colonial states took over the infrastructures, borders, and epistemologies that had previously altered the Indigenous worlds. Natural calamities, outbreaks of diseases, and climate risks—frequently made worse by colonialism's environmental degradation and resource exploitation—are still being told through the narrative of lack, which hides the Indigenous resilience and the ecological wisdom. Similarly, the Western disability discourses that came with colonization have been more applied in the Indigenous context than the local concepts of interdependence, spiritual balance, and collective responsibility have been interpreted.

However, this story is not only about grief. The Pacific peoples have created vivid counter-maps that question imperial boundaries and confirm Indigenous ways of being. The "Blue Pacific" movement, for example, sees the ocean as a route of communication instead of a divider, while Native scholars along with disability rights activists redefine physicality as a matter of relationship rather than a medical issue; Besides, the reconsideration of archives and the community-based studies bring to light the manipulations and the wiping out of certain aspects that were part of the colonial maps. This chapter ultimately argues that to understand disability in Oceania, one must acknowledge its historical production through colonial spatial, racial, and epistemic regimes. At the same time, the disability can function as a site of resistance, and people's disobedience to the colonial state's requirements for productivity, conformity, and normativity is expressed through their bodies. The chapter shifts the focus to Indigenous knowledge systems, relational geographies, and histories of survival, thus taking part in a project not only to decolonize cartography and disability studies but also to allow more just, inclusive, and culturally grounded futures to emerge.

References

Andrews, K., Parekh, J., & Peckoo, S. (2019). *A guide to incorporating a racial and ethnic equity perspective throughout the research process*. Child Trends.

Anesi, J. (2022). Enduring the storm: Dealing with mental disabilities in Oceania. *Disability Studies Quarterly, 41*(4). https://doi.org/10.18061/dsq.v41i4.8457

Barker, C., & Murray, S. (2010). Disabling postcolonialism: Global disability cultures and democratic criticism. *Journal of Literary & Cultural Disability Studies, 4*(3), 219–236. https://doi.org/10.3828/jlcds.2010.20

Bennett, C. L., Gleason, C., Scheuerman, M. K., Bigham, J. P., Guo, A., & To, A. (2021). "It's complicated": Negotiating accessibility and (mis) representation in image descriptions of race, gender, and disability. In *Proceedings of the 2021 chi conference on human factors in computing systems* (pp. 1–19). https://doi.org/10.1145/3411764.3445498

Brinkman, A. H., Rea-Sandin, G., Lund, E. M., Fitzpatrick, O. M., Gusman, M. S., & Boness, C. L. (2023). Shifting the discourse on disability: Moving to an inclusive, intersectional focus. *American Journal of Orthopsychiatry, 93*(1), 50. https://doi.org/10.1037/ort0000653

Buscher, D., & Pearce, E. (2024). Humanitarianism and disability. In *Handbook on humanitarianism and inequality* (pp. 265–279). Edward Elgar Publishing. https://doi.org/10.4337/9781802206555.00028

Castellino, J., & Keane, D. (2009). Minority rights in the Pacific region: A comparative analysis. *Oxford University Press*. https://doi.org/10.1093/acprof:oso/9780199574827.001.0001

Chand, S. (2010). *Shaping new regionalism in the Pacific Islands: Back to the future? (No. 61)*. ADB Working Paper Series on Regional Economic Integration. http://hdl.handle.net/11540/1941

Chang, D. A. (2016). *The world and all the things upon it: Native Hawaiian geographies of exploration*. University of Minnesota Press. https://doi.org/10.5749/minnesota/9780816699414.001.0001

Chape, S. (2006). Review of environmental issues in the pacific region and the role of the pacific regional environment programme. In *Workshop and symposium on Collaboration for sustainable development of the Pacific Islands: Towards effective e-learning systems on environment* (pp. 27–28). https://library.sprep.org/content/review-environmental-issues-pacific-region-and-role-pacific-regional-environment-programme

Cole, D. G., & Hart, E. R. (2021). The importance of indigenous cartography and toponymy to historical land tenure and contributions to Euro/American/Canadian cartography. *ISPRS International Journal of Geo-Information, 10*(6), 397. https://doi.org/10.3390/ijgi10060397

Cummins, R. A., & Capic, T. (2017). The history of well-being in Oceania. In *The pursuit of human well-being: The untold global history* (pp. 453–491). Springer International Publishing. https://doi.org/10.1007/978-3-319-39101-4_14

Dodd, R., Reeve, E., Sparks, E., George, A., Vivili, P., Tin, S. T. W., Buresova, D., Webster, J., & Thow, A. M. (2020). The politics of food in the Pacific: coherence and tension in regional policies on nutrition, the food environment and non-communicable diseases. *Public Health Nutrition, 23*(1), 168-180. https://doi.org/10.1017/S1368980019002118

Erevelles, N., & Erevelles, N. (2011). *Disability and difference in global contexts: Enabling a transformative body politic* (pp. 14–5). Palgrave Macmillan. https://doi.org/10.1057/9781137001184

Erevelles, N., & Minear, A. (2010). Unspeakable offenses: Untangling race and disability in discourses of intersectionality. *Journal of Literary & Cultural Disability Studies, 4*(2), 127–145. https://doi.org/10.3828/jlcds.2010.11

Fakhruddin, S. H. M., Babel, M. S., & Kawasaki, A. (2015). Assessing the vulnerability of infrastructure to climate change on the Islands of Samoa. *Natural Hazards and Earth System Sciences, 15*(6), 1343–1356. https://doi.org/10.5194/nhess-15-1343-2015

Franklin, M. I. (2006). *Postcolonial politics, the internet and everyday life: Pacific traversals online*. Routledge. https://doi.org/10.4324/9780203448991

Ganguly D, ed. (2021). Cartographic shifts. In *The Cambridge history of world literature* (pp. 409–510). Cambridge University Press. https://doi.org/10.1017/9781009064446

Gani, J. K., & Marshall, J. (2022). The impact of colonialism on policy and knowledge production in International Relations. *International Affairs, 98*(1), 5–22. https://doi.org/10.1093/ia/iiab226

Gonzales, L. (2019). Designing for intersectional, interdependent accessibility: A case study of multilingual technical content creation. *Communication Design Quarterly Review, 6*(4), 35–45. https://doi.org/10.1145/3309589.3309593

Grech, S. (2011). Recolonising debates or perpetuated coloniality? Decentring the spaces of disability, development and community in the global South. *International Journal of Inclusive Education, 15*(1), 87–100. https://doi.org/10.1080/13603116.2010.496198

Grech, S., & Soldatic, K. (2017). Introduction: Disability and colonialism:(dis) encounters and anxious intersectionalities. In *Disability and colonialism* (pp. 1–5). Routledge. https://doi.org/10.1080/13504630.2014.995394

Harley, J. B. (1989). Deconstructing the map. *Cartographica: The International Journal for Geographic Information and Geovisualization, 26*(2), 1–20. http://hdl.handle.net/2027/spo.4761530.0003.008

Harrington, C. N., Desai, A., Lewis, A., Moharana, S., Ross, A. S., & Mankoff, J. (2023). Working at the intersection of race, disability and accessibility. In *Proceedings of the 25th International ACM SIGACCESS Conference on Computers and Accessibility* (pp. 1–18). https://doi.org/10.1145/3597638.3608389

Harrington, C. N., Garg, R., Woodward, A., & Williams, D. (2022, April). "It's kind of like code-switching": Black older adults' experiences with a voice assistant for health information seeking. In *Proceedings of the 2022 CHI Conference on Human Factors in Computing Systems* (pp. 1–15). https://doi.org/10.1145/3491102.3501995

Hattori, A. P., & Samson, J. eds. (2023). The colonial era in the Pacific. In *The Cambridge history of the Pacific Ocean. The Cambridge history of the Pacific Ocean* (pp. 421–560). Cambridge University Press. https://doi.org/10.1017/9781108226875.023

Hofmann, M., Kasnitz, D., Mankoff, J., & Bennett, C. L. (2020, October). Living disability theory: Reflections on access, research, and design. In *Proceedings of the 22nd International ACM SIGACCESS Conference on Computers and Accessibility* (pp. 1–13). https://doi.org/10.1145/3373625.3416996

Jolly, M. (2007). Imagining Oceania: Indigenous and foreign representations of a sea of islands. *The Contemporary Pacific* (pp. 508–545). https://doi.org/10.1353/cp.2007.0054

Jupiter, S., Mangubhai, S., & Kingsford, R. T. (2014). Conservation of biodiversity in the Pacific Islands of Oceania: Challenges and opportunities. *Pacific Conservation Biology, 20*(2), 206–220. https://doi.org/10.1071/PC140206

Kabutaulaka, T. (2021). Mapping the Blue Pacific in a changing regional order. *The China alternative: Changing regional order in the Pacific Islands* (pp. 41–69). https://doi.org/10.22459/CA.2021.01

Leal, D. D. C., Strohmayer, A., & Krüger, M. (2021). On activism and academia: Reflecting together and sharing experiences among critical friends. In *Proceedings of the 2021 CHI Conference on Human Factors in Computing Systems* (pp. 1–18). https://doi.org/10.1145/3411764.3445263

Levering, R. B. (2016). *The cold war: A post-cold war history*. John Wiley & Sons.

Lewis, M. W., & Wigen, K. (1997). *The myth of continents: A critique of metageography*. University of California Press. http://www.nytimes.com/books/first/l/lewis-myth.html

Martinsson-Wallin, H., & Thomas, T. (2014). *Monuments and people in the Pacific*. Uppsala universitet.

Matsuda, M. K. (2006). The Pacific. *The American Historical Review, 111*(3), 758–780. https://doi.org/10.1086/ahr.111.3.758

McRuer, R. (2010). Disability nationalism in crip times. *Journal of Literary & Cultural Disability Studies, 4*(2), 163–178. https://doi.org/10.3828/JLCDS.2010.13

Meekosha, H. (2011). Decolonising disability: Thinking and acting globally. *Disability & Society, 26*(6), 667–682. https://doi.org/10.1080/09687599.2011.602860

Nijs, G., & Heylighen, A. (2015). Turning disability experience into expertise in assessing building accessibility: A contribution to articulating disability epistemology. *Alter, 9*(2), 144–156. https://doi.org/10.1016/j.alter.2014.12.001

Parekh, P. (2007). Gender, disability and the postcolonial nexus. *Wagadu: A Journal of Transnational Women's & Gender Studies, 4*(1), 13. https://digitalcommons.cortland.edu/wagadu/vol4/iss1/13

Pickles, J. (2012). *A history of spaces: Cartographic reason, mapping and the geo-coded world*. Routledge. https://doi.org/10.4324/9780203351437

Prabhakar, A. C. (2024). Colonial echoes: Unraveling Economic legacies and geopolitical shifts in the South Pacific Islands. *African and Asian Studies, 1*(aop), 1–29. https://doi.org/10.1163/15692108-bja10032

Radcliffe, S. A. (2018). Geography and indigeneity II: Critical geographies of indigenous bodily politics. *Progress in Human Geography, 42*(3), 436–445. https://doi.org/10.1177/030913251769163

Rallu, L. (2011). *What killed Pacific Islanders: Epidemics or colonization? The cases of Tahiti and the Marquesas*. [Online]. Accessed on the, 12, 2021.

Rankin, Y. A., & Thomas, J. O. (2019). Straighten up and fly right: Rethinking intersectionality in HCI research. *Interactions, 26*(6), 64–68. https://doi.org/10.1145/3363033

Rios, D., Magasi, S., Novak, C., & Harniss, M. (2016). Conducting accessible research: Including people with disabilities in public health, epidemiological, and outcomes studies. *American Journal of Public Health, 106*(12), 2137–2144. https://doi.org/10.2105/AJPH.2016.303448

Sabariego, C., Fellinghauer, C., Lee, L., Kamenov, K., Posarac, A., Bickenbach, J., Kostanjsek N, Chatterji S., & Cieza, A. (2022). Generating comprehensive functioning and disability data worldwide: Development process, data analyses strategy and reliability of the WHO and World Bank Model Disability Survey. *Archives of Public Health, 80*(1), 6. https://doi.org/10.1186/s13690-021-00769-z

Shanks, G. D. (2016). Pacific island societies destabilised by infectious diseases. *Journal of Military and Veterans Health, 24*(4), 71–74. https://jmvh.org/article/pacific-island-societies-destabilised-by-infectious-diseases/

Sharma, N. T. (2021). *Hawai'i is my haven: Race and indigeneity in the Black pacific*. Duke University Press. https://doi.org/10.1215/9781478021667

Sherry, M. (2007). (Post) colonising disability. *Wagadu: A Journal of Transnational Women's & Gender Studies, 4*(1), 14. https://digitalcommons.cortland.edu/wagadu/vol4/iss1/14

Short, J. R. (2009). *Cartographic encounters: Indigenous peoples and the exploration of the New World*. Reaktion Books. https://doi.org/10.1016/j.jhg.2010.02.001

Soldatic, K., & Biyanwila, J. (2006). Disability and development: A critical southern standpoint on able-bodied masculinity. *Annual Conference of the Australian Sociological Association* (pp. 63–76). University of Western Australia and Murdoch University.

Soldatic, K., & Gilroy, J. (2018). Intersecting indigeneity, colonialisation and disability. *Disability and the Global South, 5*(2), 1337–1343.

Soldatic, K., & Grech, S. (2014). Transnationalising disability studies: Rights, justice and impairment. *Disability Studies Quarterly, 34*(2). https://doi.org/10.18061/dsq.v34i2.4249

Trask, H. K. (1990). Politics in the Pacific Islands: Imperialism and native self-determination. *Amerasia Journal, 16*(1), 1–19. https://doi.org/10.17953/AMER.16.1.8X73270241126844

UNDESA. (2009). *State of the world's indigenous peoples*. UNDESA.

Venkataramanan, M. (2024). *The Pacific Islands: United by Ocean*. A Magazine of Ideas, Arts, And Scholarship, Public Book.

Windsong, E. A. (2018). Incorporating intersectionality into research design: An example using qualitative interviews. *International Journal of Social Research Methodology, 21*(2), 135–147. https://doi.org/10.1080/13645579.2016.1268361

Wöll, S. (2023). Unmasking maps, unmaking empire: Towards an archipelagic cartography. *Journal of Transnational American Studies, 14*(1). https://doi.org/10.5070/T814160835

Chapter 2
Policy Gaps and Global Frameworks: A Critical Review

Abstract Analyzing the interaction between the disability issue, policy-making, and disaster risk reduction (DRR) in the Pacific region, this chapter brings to focus the disconnection between global promises and local performances. Indeed, the principles of the Convention on the Rights of Persons with Disabilities (CRPD) and the Sendai Framework for Disaster Risk Reduction (2015–2030) seem to be the very foundations of the chapter. It evaluates critically how far the global mandates of inclusivity, accessibility, and participation have penetrated into the DRR policies at the national and regional levels, where traditionally only partial implementation has been the reality. It points out the need to transition from the historical perspective of disability as a welfare issue to a rights and resilience-based one, within the broader development and climate governance discourse. The chapter brings in examples from the Solomon Islands, Tuvalu, and Taiwan to substantiate its assertion that there are still gaps in policy, barriers to implementation, and challenges in accessibility, all of which block the way to a full-fledged disability-inclusive DRR in the Pacific Islands. Among other factors, these include: weak institutional capacity, poor coordination, lack of disability-disaggregated data, inadequate funding, and the continued prevalence of social stigma that is deeply rooted in society. The results, although showing some progress in the acceptance of policy, still indicate that the transition from merely symbolic inclusion to actual practice is not yet completed. The chapter discusses how participation, universal accessibility, and intersectoral collaboration form the backbone of a rights-based approach, which is vital for building just and sustainable resilience in the disaster-prone Pacific region.

Keywords Disability inclusion · Sendai framework · Disaster risk reduction · Pacific Island nations · Policy implementation · Climate change adaptation

S. M. Thomas and R. Veerabathiran, *Disability, Disaster, and Resilience in Oceania*, SpringerBriefs in Modern Perspectives on Disability Research,
https://doi.org/10.1007/978-981-95-8535-9_2

2.1 Introduction

Disability is not only a medical condition but also a complex phenomenon that interacts with factors such as society, the environment, and people's attitudes towards it. The social model of disability states that negative attitudes, inaccessible places, and societal institutions are responsible for the existence of disabilities. Similarly, susceptibility to calamities is not solely due to physical factors; rather, it is an outcome of the interaction among risk factors, institutional capacity, social inequality, and resource availability. From this viewpoint, the victories for the disabled community are being excluded from the conversation due to the framing of disability as a welfare issue or treating disaster risk as a technical problem. Hence, throughout the whole process of disaster management risk assessment, planning, preparedness, response, and recovery, mainstreaming disabilities and giving priority to accessibility, participation, and rights over optional extras will be the way to go for an effective Disaster Risk Reduction (DRR) policy (Samaha, 2007).

From this perspective, disability-inclusive disaster risk reduction, two international treaties are particularly noteworthy. The Sendai Framework (2015–2030), which has people at the center of its risk-reduction strategy, coordinates the entire governance process, always considering the needs of the most vulnerable groups, making investments that protect the environment and the global community, as well as investments in resilient recovery. What is of utmost importance is that the Research Living Framework ensures accessibility for all, allows people's unrestricted movement to safe places, and makes communication devices available for emergencies; in this way, it entails the full participation of disabled persons in the training and capacity-building, thereby integrating them into the powering up of the different processes with new capabilities and skills (Pearson & Pelling, 2015).

The Convention on the Rights of Persons with Disabilities (CRPD) provides a framework for inclusive international development and is part of the corpus of international law and treaties. Adopted in 2006, the Convention came into force in 2008. The great majority of nations in the Asia–Pacific area are among the 184 nations that have ratified the Convention as of January 5, 2022. To ensure that people with disabilities (PwD) can enjoy their rights on an equal footing with others, nations must create, modify, or enhance their national disability policies and strategies after ratifying and becoming a state party to the Convention. The CRPD Committee, a body created by the Convention to review reports from state parties and alternative reports from other stakeholders, must also receive its initial state reports. The Committee examines all submissions, consults with pertinent parties, and then releases reports, known as Concluding Observations, in response (ESCAP, 2022).

Through general remarks on relevant themes and problems, the CRPD Committee also offers recommendations to promote the creation and implementation of policies and strategies that guarantee the realization of the rights of PwD. In addition to advancing human rights law, the CRPD directs global development. In addition to offering a roadmap for how governments should change their laws, public policies, and strategies to achieve inclusive international development, it also provides a legislative framework to ensure that the rights of PwD are upheld post-ratification. Every nation's attempt to implement the Sustainable Development Goals (SDGs) (and any subsequent development plan) must incorporate the rights and principles established under the CRPD, given these dual aims (international development and the preservation and promotion of human rights) (ESCAP, 2022).

Small island developing states (SIDS), which are small, underdeveloped, and at the same time very diverse in terms of location and culture, have their own unique governance systems, face financial challenges, and suffer from environmental degradation. Besides, the situation is worsening, making disasters even more serious, as Pacific countries are already highly prone to natural phenomena such as cyclones, storm surges, and earthquakes. The Pacific region's diverse national policies reflect the different attitudes of states towards disability and disaster risk reduction. While some governments have adopted modern laws protecting the rights of PwD and have created separate DRR plans, others still apply fragmented approaches, addressing one sector at a time and with very little intra-sector cooperation (Binger et al., 2004).

Thus, this chapter presents a detailed evaluation of the regional and global policy frameworks that are related to disaster risk reduction and disability. It begins with presenting the major international pacts such as the CRPD and the Sendai Framework for DRR (2015–2030), and it underlines how those documents can be used to promote accessibility and inclusion that govern policy discussion and actions, and looks at the degree to which their objectives are incorporated (or not incorporated) into national policies in the Pacific. Subsequently, it surveys the policies that Pacific Island nations have implemented for disability and disaster management, highlighting the advancements and challenges encountered in their implementation. Moreover, the chapter delves into the socio-cultural, institutional, and infrastructural barriers that obstruct the national and local level effective implementation of these global commitments. Through the examination of the above-mentioned factors, the chapter is trying to identify the gaps in policies, difficulties in implementation, and issues in accessibility that together influence the experiences of PwD during disasters.

2.2 Disabled Inclusion in Global Frameworks: CRPD and Sendai

The integration of PwD into disaster management and global development frameworks is a clear representation of a major shift from a welfare to a rights-based approach. PwD were often excluded from participation and were only considered as beneficiaries of charity. However, the economic downturn of 2008 and the global crisis that followed created new opportunities for the empowerment of marginalized groups, including PwD. The process was also sped up by the UN CRPD ratification, which assured disabled people the right to take part in the building of inclusive, resilient societies and to be acknowledged as active changers of the situation (Njelesani et al., 2012).

The CRPD was enacted in 2006, and it was the first global treaty to cover human rights comprehensively for all kinds of disabled people. The core concepts of the Convention revolve around equality, dignity, and access; hence, it requires governments to include the rights of disabled persons in their policies and developmental programs. Moreover, the Sendai Framework is a landmark that brings about inclusive DRR by not only recognizing the concerns related to disability directly but also advocating for implementing DRR measures that are accessible and pleasant for PwD during the whole process, including their active participation (Schulze, 2010).

Both frameworks are in full agreement on one critical aspect: the notion of resilience itself is impossible to even consider without everyone's involvement. They argue that the exclusion of disabled individuals is not only a violation of the basic social contract but also reduces the capacity of a community to cope with disasters and maintain itself. By aligning the fundamental rights of disabled people with the action-oriented needs of the Sendai Framework, a global policy is beginning to narrow the gap between human rights promotion and practical disaster management. The overlap of policies is evidence of a growing recognition that the participation of disabled people strengthens society as a whole by making governance systems more resilient, ensuring a fair distribution of resources, and improving the community's ability to respond to crises. The CRPD and SFDRR, combined, are therefore the policymakers' guide to the construction of a truly global, inclusive framework where no one is left behind, either to witness or to benefit from development or disaster.

2.2.1 *People with Disabilities' Rights and the CRPD Policy*

Inclusion of PwD in global human rights discussions is from an ultra-basic right of people perspective, placing their rights as the primary point across the different aspects of life. In the past, persons with disabilities were seen mainly as a problem needing a medical solution or as candidates for charity work, which often resulted in their exclusion from any policy discussions and in their being denied opportunities to participate in social, economic, and political spheres. The United Nations' adoption of the CRPD in 2006 changed this scenario by characterizing people with disabilities as active participants in society and equal citizens. The Convention imposes an elaborate set of legal duties on the State Parties not only to advance but also to protect and secure the full and equal enjoyment of all human rights by disabled persons. Besides, it lays down principles such as non-discrimination, access, participation, and the recognition of the innate dignity of every individual, while requiring governments to change laws, institutions, and policies that keep people apart.

The CRPD has, through this framework, emerged as a pivotal factor in the global fight for the creation of inclusive communities, the realization of social justice, and the empowerment of the disabled population. It enforces the aforementioned non-discrimination, accessibility, participation, and respect for inherent dignity as the values that must be at the forefront of the campaign while compelling the governments to implement the necessary legal, institutional, and policy changes that will eradicate exclusion. By means of this strategy, the CRPD has turned into the cornerstone for worldwide projects aimed at encouraging the development of inclusiveness, social justice, and liberation of the disabled across the globe (Assembly, 2011; De Beco, 2019).

The term "climate-resilient development" emphasizes the requirement for a completely innovative method of development that will guarantee the survival of both the human race and nature. In a more general sense, it means the development plans that try to achieve the sustainable development goal, together with the adaptation to climate change, vulnerability reduction, and greenhouse gas emissions control. It has emerged as a key phrase in the Intergovernmental Panel on Climate Change (IPCC) evaluation process, which offers a scientific foundation for global climate policy. The universally accepted normative SDGs frequently guide climate-resilient development. In a nutshell, climate-resilient development pathways (CRDPs) aim to guide development to ensure better welfare, social justice, and the health of people and the planet (Roy et al., 2018).

Four essential components of climate-resilient development can be better understood through the CRPD and lived experiences of disability: social justice and equity as normative objectives; the moral foundations of social decisions; the unequal relationships that lead to marginalization; and how society negotiates uncertainty through inclusive and contestatory politics. In addition to clarifying how marginalization creates vulnerability, a disability lens also helps elaborate on the ethic of solidarity as the foundation for social decisions and a guide to development towards climate-resilient paths. In both official and informal decision-making processes, social justice includes non-discrimination and fair participation. Through their understanding of human connection and of how to solve problems in the midst of uncertainty, people with disabilities' resilience knowledge contributes to a rethinking of sustainable development (Eriksen et al., 2021).

In fact, the COVID-19 epidemic measures of the past year and a half have made people realize how essential aspects of our daily lives, such as social connection, a sense of belonging, solidarity, and equity, are. Self-actualization, affective bonds, interpersonal relationships with friends and family, pleasure, satisfaction, self-efficacy, growth, purpose, mastery, meaning, harmony, awareness, social position, and influence on one's life and surroundings are just a few of the intangible aspects of well-being, which also includes meeting standards of living, health, physical safety, work, and leisure needs (Carlquist et al., 2017; Næss et al., 2011). On the other hand, social injustices are both procedural and distributional. Not only do COVID-19 and climate change negatively affect the well-being of already marginalized groups, like children and PwD, but they are also frequently glaringly underrepresented or completely excluded from decision-making processes that deal with these issues (Orsander et al., 2020; Pineda & Corburn, 2020).

Significant intellectual functioning limitations (IQ < 75) and restrictions in adaptive behavior across three skill areas (conceptual, social, and practical) are characteristics of intellectual disability (ID), which first manifests before age 18. Services from the health, education, and social welfare sectors offer the specialized, integrated therapy that people with ID need. Every person should have the right to legal capacity in accordance with Article 12 of the UNCRPD. People need to be able to make their own decisions and share them with others to exercise this fundamental human right. To do this, services must use assisted decision-making techniques, which would fundamentally alter how families, professionals, service providers, and the general public view and interact with people with ID (Werner, 2012).

In the healthcare, social services, and educational areas, the support needed for people with ID through the proxy decision-making process is clearly visible. This is a complex process with several important steps. To begin with, in the modern world, people with ID should be provided with impartial education and training that equip them with the skills to make informed choices. Moreover, these people should be taught about setting objectives, self-control, and feeling competent through one-to-one tutoring in a manner that best fits their mental and learning skills. The second step is to move the service provision towards a person-centered approach in order to provide an authentic choice and way of life that is flexible according to the individual's daily activities and likes. This is achieved by encouraging self-advocacy, consumer-directed services, and active participation in decision-making. Third, caregivers, both family and professional, need to be trained in effective communication and supportive practices that enable understanding and autonomy. They should employ techniques such as visual aids, focus on the decision-making process rather than the outcomes, and evaluate decision-making capacity on an individual basis. Acknowledging that the process of making mistakes is part of learning, caregivers must be willing to have discussions about sensitive matters with the proper preparation and with much understanding (Brown & Brown, 2009; Dukes & McGuire, 2009; Ferguson et al., 2011; Gruskin, 2002; Heller et al., 2011; Holburn & Cea, 2007).

Therefore, the CRPD can be used to promote change that will improve the rights of PwD. There was a time when states may have interpreted the right to work in a way that would either increase or reduce social inclusion. This ambiguity has been eliminated by the CRPD, which outlines specific measures governments must take to ensure that PwD can exercise their right to work. Moving ahead, the implementation problem becomes the focus. Before 2006, policies about people with disabilities lacked explicit, objective standards by which to be evaluated. Since its inception, the CRPD has served as a benchmark for good governance and a catalyst for change, allowing organizations that support disabled people and scholarship on their rights to place greater emphasis on implementation. Existing rights are re-articulated in the CRPD, which also clarifies how those rights should be realized for people with disabilities and gives organizations representing those individuals a say in how the Convention is implemented. Scholars and campaigners for disability rights now have the responsibility to realize the CRPD's potential and use it as a tool to push for reform (Harpur, 2012).

2.2.2 *The Sendai Framework and People with Disability*

The Yokohama Strategy and Plan of Action for a Safer World: Guidelines for Natural Disaster Prevention, Preparedness, and Mitigation was the result of the inaugural World Conference on Disaster Risk Reduction (WCDRR), which was held in Yokohama, Japan, in 1994 (UN, 1994). There were no allusions to individuals with disabilities, nor were there any references to disability-related vocabulary or concepts such as universal design, accessibility, or inclusion in the Yokohama Strategy document. Vulnerability was mentioned in a few places in the paper, but only in relation to developing nations and not to specific groups of people. Vulnerable communities and groups, and an inclusive approach to disaster risk reduction, were not included until the Review of the Yokohama Strategy and Plan of Action for a Safer World (UNISDR, 2004), published following the discussions in Hyogo, Japan, in January 2005 (Tozier & Baudoin, 2015).

Building the Resilience of Nations and Communities to Disasters (HFA) is the Hyogo Framework for Action 2005–2015, developed at the second WCDRR in Hyogo, Japan, in 2005 (UNISDR, 2005). Once more, there was no explicit reference to those with disabilities as a vulnerable category. Only within the framework of catastrophe, education and training, and gendered perspectives curricula was the subject of inclusion discussed. There was no discussion of the dangers individuals with impairments face in the HFA document. Explicit suggestions for a disability-inclusive disaster risk-reduction framework and its implementation were included at the Third UN WCDRR 2015 in Sendai, Japan (FEMA, 2015). Both attendees and presenters with impairments may access the third WCDRR location and conference sessions. At central locations, closed captioning was available in both English and Japanese, and sign language interpretation was available upon request for different sessions. The venues offered wheelchair-accessible transportation. Participants who were blind were given devices that displayed Braille, and the documents were in an accessible format. Of paramount importance, almost 200 individuals with disabilities actively engaged in the WCDRR proceedings as authors, presenters, panelists, or delegates (Stough & Kang, 2015).

The crucial role that people with disabilities and disability advocacy groups play as partners and stakeholders in emergency preparation and recovery is the last important subject. One of the main goals of the disability community has long been inclusion in communal life. The SFDRR recognizes individuals with disabilities as potential active participants alongside the rest of the community, treating them as stakeholders in the planning and execution of disaster risk reduction. "In a survey of people living with disabilities that we conducted last year, a large majority told us that they want to be consulted about their needs to face and prepare for disasters, as well as be able to contribute expertise and participate in planning and implementation," said Margareta Wahlström, Head of the UN Office for Disaster Risk Reduction. It is important to remember that impairment does not equate to incapacity (UNISDR, 2013). Other vulnerable groups may benefit from actions performed on behalf of individuals with disabilities. Other vulnerable groups are empowered to become active participants in disaster risk reduction, including individuals with disabilities and their organizations. The SFDRR now includes aspects it did not before the inclusion of disability and disability-related issues. Everyone can benefit from the ideas of universal design, accessibility, and inclusion, not only those with impairments. All things considered, including individuals with disabilities in planning and policy not only makes them safer, but also somewhat protects everyone. As part of the SFDRR process and the final product, the Disability Caucus successfully pushed for commendable policy reforms (Stough & Kang, 2015).

2.3 Pacific Island Nations and National Disability and Disaster Risk Policies

The Pacific islands are small in size and thus bear the brunt of natural disasters like floods, cyclones, storm surges, volcanic eruptions, earthquakes, and tsunamis, which happen regularly in the area, and the Economic and Social Commission for Asia and the Pacific has referred to this region as the most disaster-prone area in the world (ESCAP, 2015, 2016). Most countries in the Pacific region are experiencing economic instability due to limited resources, small size, and remoteness (World Bank, 2018).

In 2005, during the 36th Pacific Islands Forum, leaders of the Pacific Islands gave their seal of approval to the Pacific Islands Framework for Action on Climate Change 2006–2015 (PIFACC). The period of 2006–2015, as the framework's time, was consistent with the Millennium Declaration, the Johannesburg Plan of Implementation, and the remaining activities of the UN Commission on Sustainable Development. A Pacific Islands Climate Change Roundtable (PCCR) conference in 2005 reviewed the framework. One outcome of the review was the creation of an action plan for implementing the framework. The Pacific Regional Environment Programme (SPREP) was asked in 2008 to convene regular PCCR meetings. The Development Partners for Climate Change (DPCC), which comprises Suva-based bilateral and multilateral donor organizations and allied organizations, also convenes regularly to coordinate climate change-related efforts across the Pacific (UNISDR, 2012).

Climate change is a hazard in the Solomon Islands through rising sea levels and increased hydro-meteorological occurrences. With 14% of people living under the poverty line, and more than a third of infants being affected by a lack of growth, it is one of the poorest Pacific countries, ranked with the lowest per capita income. The diverse ethnic backgrounds of the country's around 500,000 people are affected by poverty, inequality, unstable politics, and weak institutions. It is estimated that about 80% of the population depends on subsistence agriculture and lives in rural areas. Hence, the Solomon Islands are ranked sixth among 171 countries in the World Risk Index as the most prone and affected by natural disaster risks. Among the significant catastrophes that have taken place not long ago and produced such an effect are the 2007 tsunami that claimed 52 lives, 10,000 injured, and caused

US$91 million damage, and the 2014 Honuara flash floods, which caused 22 deaths, displaced 11,000 people, and resulted in US$107.8 million losses (King et al., 2019).

As per the Solomon Islands National Policy on Disability (2005–2010), a disability is a condition that prevents "normal ways of living." Prevalence estimates, however, vary from 1.66% to 16% due to inconsistent definitions and data collection methods (King et al., 2019). The 2009 census revealed that 14% of the population had mobility, memory, hearing, or visual impairments (Gibson et al., 2015). In the Solomon Islands, people with disabilities are still very much the most disadvantaged group in society. They are not able to get education, jobs, or access to services like everybody else, and this is contrary to a global disability occurrence of about 15%. Persons with disabilities are often subjected to the stigmas of traditional beliefs that regard them as having been cursed or punished and thus are not only pushed to the periphery of society but also rendered invisible (Gartrell et al., 2018).

PwD receive disproportionate support during disasters, have very little access to information, and are rarely included in disaster management planning. Through their efforts, disability rights organizations and NGOs have managed to make some progress even in the face of stigma. Their actions were supported by international and regional frameworks like the UN Convention on the Rights of PwD, the Incheon Strategy (2013–2020), and the Sendai Framework (2015–2030), which have all strengthened the government's determination to promote inclusiveness (Twigg et al., 2018).

Article 11 of the UNCRPD (2006), which the Solomon Islands signed but did not ratify, stipulates that PWD must be taken care of in times of disaster. In the Incheon Strategy, developed by ESCAP with the participation of Pacific governments and civil society, the focus was on the need to include PwD in disaster management. The Sendai Framework, one of the major international agreements of 2015, mandates that PWD be part of risk assessment, planning, and recovery. Despite the support, the Pacific area has made very little progress towards disaster management that includes the disabled singing industry; thus, it is necessary to take deliberate actions to remove the existing discrimination (King et al., 2019; Motz, 2018).

Disaster governance in Taiwan, in the Asia–Pacific region, for the disabled and other PWD is found to primarily apply the Sendai Framework principles in emergency response and preparedness, while providing little in the recovery and rebuilding phase. The main tactic during the preparedness phase is to create a list of PWD and the social welfare services available in potential areas of catastrophe. However, this approach focuses on identifying people instead of establishing contact channels to understand their needs before, during, and after a disaster. Welfare services hardly collaborate, and there are very few opportunities to exchange experiences regarding disasters. Public hearings, such as those on power backup plans, have been held only once (Lee & Chen, 2019).

The first strategy in the emergency response phase is the pre-emptive evacuation by means of rehabilitation buses, which, during drills, is a common practice. Nonetheless, there are only a limited number of shelter operations for people with disabilities, and few evaluations exist on whether facilities or shelters could accommodate them. Moreover, there is no evidence that persons with disabilities took part in the drafting of shelter or evacuation strategies. More projects have been undertaken to improve early warning systems, such as flood or mudslide phone notifications issued by different government agencies, the creation of LINE chat groups for facilities, and the provision of sign language translations for typhoon announcements. The third priority of the Sendai Framework, investment in disaster risk reduction for resilience, together with priorities one (understanding the risk) and two (governance strengthening), is primarily aligned with most ongoing initiatives. This only reflects at the policy level and not at the implementation level, the slight Priority 4 (readiness and recovery improvement) engagement (Lee & Chen, 2019).

Tuvalu is among many Pacific Island nations that are highly vulnerable to climate change disasters such as cyclones, floods, and rising sea levels. The country's small population, lack of basic services, and difficult location make it harder for the government to plan for and respond to disasters, and even more so for PWDs. Tuvalu's involvement in international accords like the Sendai Framework for Disaster Risk Reduction (2015–2030) and the UNCRPD has been a major reason why disability inclusion has slowly but surely gained recognition. These policies not only recognize the rights, participation, and safety of PWD during disasters but also align with the national goal of empowering through inclusiveness, government, and community resilience (Elisala et al., 2020; Handmer et al., 2025).

The National Strategic Action Plan (NSAP) for Climate Change and Disaster Risk Management of Tuvalu has included disability concerns in its agenda by acknowledging the importance of improving the resilience of the at-risk groups. Unfortunately, the unavailability of local data on PWD and the lack of resources are major hindrances to implementing the plan. The safety of the general public is often the primary concern of community-level emergency preparedness initiatives, which have very few measures to ensure accessibility for PWD, their communication, and their participation in planning and response actions. Meanwhile, the participation of PWD in exercises, communication systems, and decision-making is a work in progress, and the NGOs' and regional partners' advocacy has been a significant factor in making the problem known and promoting inclusive disaster risk reduction (UNISDR, 2012). Tuvalu's policy frameworks express a symbolic commitment to inclusion. Notwithstanding the operative gaps, these promises are still to be realized. Among the challenges are the absence of disability-disaggregated data, the lack of interagency coordination, and the inadequate supply of disability training for emergency responders. To enable disabled persons' inclusive disaster management, Tuvalu has to develop capacity, provide communication infrastructure that is accessible, and invite PWD to participate actively in decision-making at the community and government levels. Adopting a rights-based and durable strategy in line with the Sendai targets would require much more collaboration between national authorities, disability organizations, and international partners (Elisala et al., 2020).

Pacific Island nations like Tuvalu and the Solomon Islands, as well as other Asia–Pacific nations, Taiwan included, are increasingly realizing that disaster risk governance should not exclude people with disabilities. However, there are still a great deal of obstacles to overcome before these commitments made in policies will actually be put into practice. One of the turning points that advocates the rights and involvement

Table 2.1 Policy and implementation gaps in Pacific Island disaster governance

Category	Barrier	Impact on disability inclusion
Institutional weakness	Weak coordination, inadequate training, and poor understanding of the root causes of vulnerability	PwD excluded from risk assessment, planning, and recovery processes
Financial constraints	Inclusive DRR initiatives depend on short-term donor funding	No sustained investment in accessible shelters, data, or capacity-building
Accessibility failures	Inaccessible shelters, lack of ramps/toilets, inaccessible early warnings, limited sign language or alternative formats	PwD face higher risks during evacuation, communication breakdowns, and limited recovery access
Data gaps	Lack of disability-disaggregated data in Solomon Islands, Tuvalu, Vanuatu	Poor planning and resource allocation; PwD remain invisible in policy
Participation barriers	PwD treated as passive beneficiaries rather than active agents	Loss of local knowledge, needs misrepresented, weak accountability

of People with Disabilities (PWD) in every step of disaster management is the harmonization of national policies with international norms, including the UNCRPD, the Incheon Strategy, and the Sendai Framework for Disaster Risk Reduction. However, the effective participation of persons with disabilities still suffers from the impact of the stigma attached to them, a lack of proper and reliable disability-disaggregated data, poor financial support, and a lack of cooperation between the different agencies dealing with these issues. In order to provide a fair share of resilience, it is important to build up the institutions, involve the whole community, and make access to infrastructure, communication, as well as emergency response systems truly universal for all. Pacific Island nations will need to embark on a journey towards an effective disaster management system, where no one is left out, by moving from just ceremonial policy backing to real practice, grounded in strong, united, and open support. The policy and the implementation gaps in the Pacific Island Disaster Governance are illustrated in Table 2.1.

2.4 Implementation Gaps and Accessibility Challenges

2.4.1 Adoption and Operationalization: The Difference Between Policy Pledges and Practice

Pacific Island nations have not only shown a strong international commitment but also made significant political efforts to introduce universal disability and DRR measures. They are both the CRPD and the Sendai Framework, which state that PWD should be part of the whole process of disaster risk management—from assessing risks, giving warnings, and being prepared to responding and recovering to rebuilding and reconstruction (ESCAP, 2022).

As an illustration, the Pacific Disability-Inclusive Humanitarian and Resilient Development Strategy 2025–2035 not only explicitly cites Article 11 of the CRPD but also the Sendai Framework's call for inclusive DRR. Nevertheless, various studies point to the same issue: despite the existence of frameworks, implementation is slow. The 2023 thematic report on disability inclusion in the Pacific points out that the "policy-implementation" gap persists, and many commitments remain symbolic rather than operational. The review by the Stockholm Environment Institute in the Asia–Pacific region identified three significant barriers: (a) the understanding of root causes of vulnerability (including disability) is still not adequate; (b) there is no stable funding for inclusive DRR efforts; (c) poor coordination among stakeholders results in duplicating or missing out on disability-specific actions. Thus, in the Pacific context, the national disaster risk management plans are adapting to the extent that they are increasingly acknowledging "persons with disabilities", yet the conversion into accessible shelters, inclusive early warning systems, evacuation plans, and recovery strategies is still not uniformly applied (Forlin et al., 2015; Lee & Chen, 2019; Mwendwa et al., 2009; Numbers, 2023).

2.4.2 Accessibility Challenges: Early Warning, Infrastructure, Communication

Accessibility, while being the central aspect of inclusive disaster management, remains a problem in the Pacific Islands in various ways. The situation of the Pacific islands has a lot to do with the fact that most of the time the early warning and evacuation messages are not only inaccessible to the disabled persons but they are also communicated through channels that are not visible to them, due to the lack of expert sign language interpreters, inadequate format conversion (for example, visual, audio, tactile), or community-based dissemination not being offered to those who have mobility, sensory, or cognitive impairments. For example, the Global Facility for Disaster Reduction and Recovery (GFDRR) report focusing on the Pacific region, points out that evacuation plans might neglect specific requirements of a group consisting of persons with disabilities, communication routes might not connect with far-off islands, or connections with people having sensory disabilities could be ineffective, and finally, there is no accessible infrastructure (Berry & Jennette, 2022; World Bank Group, 2022).

Physical infrastructure and shelters: The circumstances are such that very few Pacific community evacuation centers and shelters are either constructed or remodeled to accommodate the needs of disabled persons with mobility or sensory impairments. A recent study conducted in the Fiji Islands revealed that although the Community-Based Disaster Risk Management (CB-DRM) policies include people with disabilities in their plans, the shelters take away their chances to find shelter due to lack of ramps, accessible toilets, tactile/visual signage, or orientation for people with disabilities; and the training of local responders does not even touch on disability-sensitive procedures most of the time (Keen et al., 2022; Nand et al., 2024).

Communication and information: Many Pacific countries lack a unified approach to collecting disability-disaggregated data or to providing disaster risk information in accessible formats. The thematic report on the Pacific points out that there is limited collection and use of disability data, leading to poor planning and inadequate funding for inclusive DRR. Vanuatu is one such country where, in the aftermath of Cyclone Pam, children and women with disabilities were found to have the least access to information about evacuation and safe shelters, the problem being more pronounced in the case of those with sensory and mobility impairments. Therefore, accessibility problems are not just a matter of infrastructure; they also involve communication, data, training, and resource allocation (Bennett, 2020; UNISDR, 2015; Strategy).

2.4.3 *Participation and Voice: Persons with Disabilities as Actors—Not just Beneficiaries*

One of the main points in Pacific DRR discourse is the very little to no participation of PWD and their representative organizations in disaster management, sometimes even at the national, local, and community levels. Global commitments (e.g., Sendai Framework's "all-of-society" approach) not just recognize but also empower PWD. However, studies suggest that PWD are mostly regarded as passive aid recipients rather than as active participants in the planning, implementation, or monitoring processes. For instance, the SEI/UN Women review noted that many DRR strategies label PWD as "passive agents of aid" rather than "agents of change". The Pacific thematic report highlights that "the leadership of PWD is critical to successful, local decision-making in DRR and should be further promoted." The community-level DRR projects in Fiji practice the "inclusiveness" label that is usually applied top-down: policies might talk about disability inclusion but in real life PWD are not fully represented in community disaster committees, trainers, and drills (Guernsey & Scherrer, 2017; Izutsu et al., 2019; Keen et al., 2022). When there is no real participation, the particular insights, preferences, and needs of PWD (which vary widely by impairment type, location, and socio-economic status) are pushed to the margins. A more significant issue, when DRR stands on human rights, is the loss of effectiveness resulting from closing the door of participation.

2.4.4 Coordination, Institutional, and Resource Limitations

The Pacific's DRR implementation experienced a range of difficulties, among which were inadequate financial support, ineffectiveness of governmental institutions, lack of cooperation, and limited oversight. These structural issues have made it very difficult for DRR practices to be inclusive. Budgeting and finance: Dedicated funding and a commitment to long-term funding are prerequisites for inclusive DRR initiatives (like accessible shelters, assistive technology, and specialist training). According to the SEI/UN Women Report, the majority of inclusive DRR initiatives in the Asia–Pacific region are supported by donor-driven, time-limited projects rather than by a continuous domestic support system. This is why the Pacific report reiterates that budget allocations for disability in risk-reduction projects remain weak. Institutional capacity and training: Responders, disaster managers, and local officials often lack the required training in disability-inclusive practices (e.g., assisting with device use, accessible evacuations, and communicating with people with sensory impairments) (UNISDR, 2012). Without training and professional development, policy statements cannot be transformed into actions. For instance, the sensitization of disaster relief operations towards persons with disabilities in the Sri Lankan context showed that, despite the existence of guidelines, local responders might disregard the voices of PWD and grass-roots workers. Coordination and data: Disability is commonly managed as a separate issue by the social welfare or health sector, while DRR is handled by disaster management agencies, resulting in isolated responses. The thematic report on the Pacific advocates the successful establishment of a multi-stakeholder approach and standardized data collection as the way to go. When there is no coordination, duplication, gaps, and inconsistency in accessibility becomes more probable (Gunarathna & Premarathne, 2024).

Monitoring, evaluation, and accountability: Policy conversion into several measurable outcomes necessitates diverse indicators, starting-point data, and continuous observation. The nonexistence of disaggregated data for persons with disabilities is constantly pointed out as an obstacle in the Pacific region: "the gathering and usage of disability data to comprehend better vulnerability and disaster risk" is one of the significant suggestions. In the absence of that, it is hard to track improvements, make necessary changes to the implementation, and ensure that the responsible parties are held accountable (Chakrabarti, 2021; ESCAP, 2018; Farid et al., 2024).

2.4.5 Recovery, Reconstruction, and Long-Term Resilience: The "Blind Spot."

Disability-inclusive practice across the phases of recovery and reconstruction, rehabilitation, and long-term resilience is largely neglected, as DRR gives much attention to preparedness and response. The Sendai Framework highlights the importance of inclusive disaster recovery and reconstruction. The various accessibility challenges in the disaster cycle are depicted in Fig. 2.1.

However, evidence from the Pacific shows that this is a significant gap. According to the thematic report, the facilities for recovery and reconstruction to be inclusive are still "limited." For instance, shelters constructed after a disaster may lack proper accessibility features unless PwD are involved in the reconstruction planning, and so forth. The Fiji case study indicates that while DRR is referenced in policies, its actual inclusion during the recovery stage (e.g., infrastructure for PwD, livelihood recovery for PwD, restoration of assistive devices) is very weak. Accessibility problems (mobility, communication, support networks) that PwD experience during disasters persist after the immediate disaster event and make it even harder for them to recover during the prolonged recovery phases, yet recovery plans seldom disaggregate by disability. This results in PwD being left behind in the longer-term

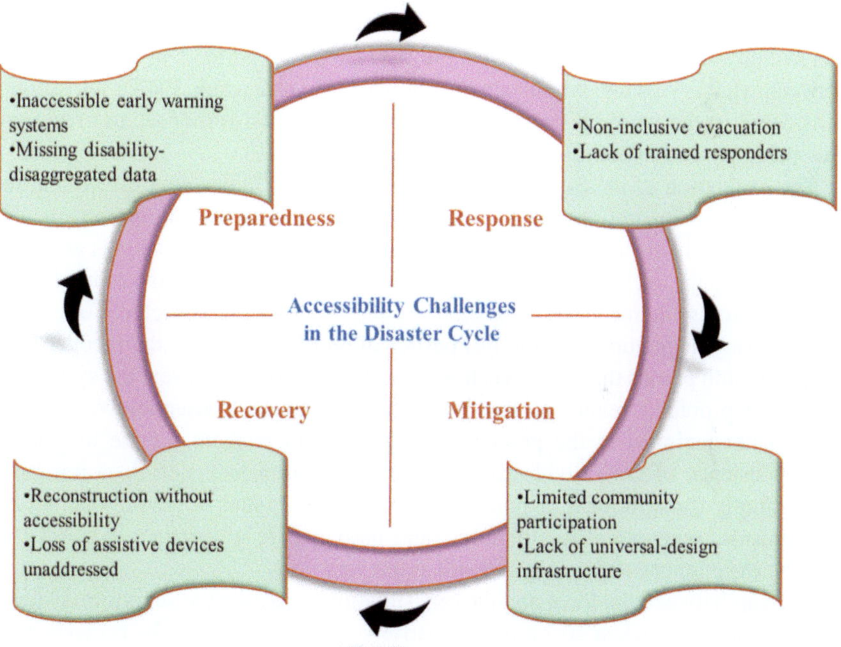

Fig. 2.1 Accessibility challenges across the disaster cycle. This figure illustrates the key accessibility challenges experienced by persons with disabilities throughout the disaster management cycle. During the preparedness phase, barriers arise from inaccessible early warning systems and the absence of disability-disaggregated data. In the response phase, non-inclusive evacuation procedures and a shortage of trained responders create significant risks. The mitigation phase is marked by limited participation of persons with disabilities in planning processes and the lack of universal-design infrastructure. In the recovery phase, reconstruction efforts often overlook accessibility needs, and the loss of assistive devices frequently remains unaddressed

resilience-building process (Spurway & Griffiths, 2016; Stjernholm, 2024; Stough & Kang, 2015).

2.5 Conclusion

The Pacific region is characterized by a variety of factors, including the entire region being the most disaster-prone, socio-economic problems, and political unrest. Besides that, the issue of disabled people taking part in disaster risk management remains a big problem. Although there are formal agreements with the CRPD, the Sendai Framework, and the Incheon Strategy, disability inclusion in disaster risk reduction is still poorly implemented in most Pacific Island countries, with outcomes mostly symbolic. The studies show that, on the one hand, the burden of disability in terms of deaths is less than in the case of others, and there is still no recognition of them among the different stakeholders; however, on the other hand, the policies increasingly see disability and the people who have it as partners in DRR. The gap between what is said and what is done continues, and as a result, PWDs are still not considered in the whole process of disaster preparedness, response, and recovery, thus being pushed to the margins.

The evaluation of the Solomon Islands, Tuvalu, and Taiwan has shown that the combination of institutional and societal barriers, such as the stigma attached to PwD and poor inter-agency coordination, together with the lack of funds and data, contributes to the exclusion of PwD. Accessibility issues are still a major hindrance; early warning systems, evacuation shelters, and recovery infrastructures often lack the necessary adaptations for the full participation of PwD. Also, the recovery and reconstruction phases that DRR has labeled as the "blind spot" reveal the systematic neglect of problems related to the disability community. Consequently, the PwD are still not included in the process of making them resilient for the long term. The persistence of such divides indicates that disability inclusion does not come about simply through legal or policy reasons—it is a matter of having operational mechanisms, accountability structures, and participative governance systems that take the experiences and voices of PwD at the very core.

The gap between the ideal and the real situation regarding the support/assistance and the measures taken to eliminate barriers must be completely bridged by a full range of measures that comprise: institution-building, disability data systems by group, long-term funding, and community-based approaches that enable and empower rather than create dependency. Moreover, governments should stop the top-down, donor-driven approach and, instead, adopt rights-based frameworks that are localized and allow the full integration of disability in all disaster management phases. The same applies to the involvement of DPOs, civil society, and regional organizations like SPREP and ESCAP, which will be the main source of inclusive resilience strategies that cater to the Pacific context.

The idea of implementing disability-inclusive disaster governance in the Pacific Ocean is not just a matter of law and regulations, but rather an opportunity for moral and human development, according to the chapter. The idea of resilience is encapsulated in the fact that, in addition to the prepared buildings and facilities, every person, no matter their physical, sensory, or cognitive ability, can be involved, helped, and respected through the giving and taking of help. The move from tokenism to real commitment will determine whether the Pacific islands can live up to the Sendai Framework's central principle: "Leave no one behind."

References

Assembly, G. (2011). Report of the Committee on the Rights of Persons with Disabilities. First Session (23–27 February 2009), Second Session (19–23 October 2009), Third Session (22–26 February 2010), Fourth Session (4–8 October 2010). https://www.bayefsky.com/general/a_66_55_2011.pdf

Bennett, D. (2020). Five years later: Assessing the implementation of the four priorities of the Sendai framework for inclusion of people with disabilities. *International Journal of Disaster Risk Science, 11*(2), 155–166. https://doi.org/10.1007/s13753-020-00267-w

Berry, A., & Jennette, L. (2022). *Disability inclusion in disaster risk management-assessment in the Caribbean region.* https://www.preventionweb.net/publication/disability-inclusion-disaster-risk-management-assessment-caribbean-region

Binger, A., Khaka, E., & Lavieren, H. V. (2004). *UNEP and Small Island developing states.* https://www.um.edu.mt/library/oar/bitstream/123456789/61275/1/UNEP_and_Small_Island_developing_states.pdf

Brown, I., & Brown, R. I. (2009). Choice as an aspect of quality of life for people with intellectual disabilities. *Journal of Policy and Practice in Intellectual Disabilities, 6*(1), 11–18. https://doi.org/10.1111/j.1741-1130.2008.00198.x

Carlquist, E., Ulleberg, P., Delle Fave, A., Nafstad, H. E., & Blakar, R. M. (2017). Everyday understandings of happiness, good life, and satisfaction: Three different facets of well-being. *Applied Research in Quality of Life, 12*(2), 481–505. https://doi.org/10.1007/s11482-016-9472-9

Chakrabarti, D. (2021). *Mainstreaming disaster risk reduction for sustainable development: A guidebook for the Asia-Pacific.* United Nations Economic and Social Commission for Asia and the Pacific (UNESCAP).

De Beco, G. (2019). The indivisibility of human rights and the convention on the rights of persons with disabilities. *International & Comparative Law Quarterly, 68*(1), 141–160. https://doi.org/10.1017/S0020589318000386

Dukes, E., & McGuire, B. E. (2009). Enhancing capacity to make sexuality-related decisions in people with an intellectual disability. *Journal of Intellectual Disability Research, 53*(8), 727–734. https://doi.org/10.1111/j.1365-2788.2009.01186.x

Elisala, N., Turagabeci, A., Mohammadnezhad, M., & Mangum, T. (2020). Exploring persons with disabilities preparedness, perceptions and experiences of disasters in Tuvalu. *PLoS ONE, 15*(10), Article e0241180. https://doi.org/10.1371/journal.pone.0241180

Eriksen, S. H., Grøndahl, R., & Sæbønes, A. M. (2021). On CRDPs and CRPD: Why the rights of people with disabilities are crucial for understanding climate-resilient development pathways. *The Lancet Planetary Health, 5*(12), e929–e939. https://doi.org/10.1016/S2542-5196(21)00233-3

ESCAP, U. (2015). *Disasters in Asia and the Pacific: 2015 year in review.* https://www.unescap.org/resources/disasters-asia-and-pacific-2015-year-review

ESCAP, U. (2018). *Disaster risk financing: opportunities for regional cooperation in Asia and the Pacific.* https://repository.unescap.org/items/f0a1b4f9-edb3-4aef-8e56-d5b28ad6098c

ESCAP, U. (2022). *Framework for disability policies and strategies in Asia and the Pacific*. https://repository.unescap.org/server/api/core/bitstreams/b0b7e59b-6e46-43e9-9167-4320b7c5f649/content

ESCAP. (2016). Coping with natural disasters in the Pacific, no 35. Suva, Fiji: UN. https://www.unescap.org/resources/mpfd-policy-briefs-no35-april-2016-coping-natural-disasters-pacific

Farid, Z. I., Islam, M. A., Roberts, P. S., & Glick, J. (2024). *Disability Inclusive Disaster Risk Reduction (DIDDR) in South Asia: Status, prospects, and challenges*. https://hdl.handle.net/10919/124180

FEMA (Federal Emergency Management Agency). (2015). *FEMA external affairs bulletin week of March 30, 2015*. http://content.govdelivery.com/accounts/USDHSFEMA/bulletins/fc0afc.

Ferguson, M., Jarrett, D., & Terras, M. (2011). Inclusion and healthcare choices: The experiences of adults with learning disabilities. *British Journal of Learning Disabilities, 39*(1), 73–83. https://doi.org/10.1111/j.1468-3156.2010.00620.x

Forlin, C., Sharma, U., Loreman, T., & Sprunt, B. (2015). Developing disability-inclusive indicators in the Pacific Islands. *Prospects, 45*(2), 197–211. https://doi.org/10.1007/s11125-015-9345-2

Gartrell, A., Jennaway, M., Manderson, L., Fangalasuu, J., & Dolaiano, S. (2018). Social determinants of disability-based disadvantage in Solomon islands. *Health Promotion International, 33*(2), 250–260. https://doi.org/10.1093/heapro/daw071

Gibson, J., Simler, K., & Carnahan, M. (2015). *Solomon Islands poverty profile based on the 2012/13 household income and expenditure survey*. Honiara Solomon Islands National Statistic Office. https://documents.worldbank.org/en/publication/documents-reports/documentdetail/922811528186449003

Gruskin, S. (2002). Ethics, human rights, and public health. *American Journal of Public Health, 92*(5), 698–698. https://doi.org/10.2105/ajph.92.5.698

Guernsey, K., & Scherrer, V. (2017). *Disability inclusion in disaster risk management: promising practices and opportunities for enhanced engagement*. Washington DC. https://www.preventionweb.net/publication/disability-inclusion-disaster-risk-management-promising-practices-and-opportunities

Gunarathna, I., & Premarathne, P. (2024). Sensitisation of disaster relief operations towards persons with disabilities. In *Climate-related human mobility in asia and the pacific: Interdisciplinary rights-based approaches* (pp. 51–68). Springer Nature Singapore. https://doi.org/10.1007/978-981-97-3234-0

Handmer, J., Monson, R., & Schinko, T. (2025). Addressing the diversity of Loss and damage in Pacific Island countries to foster a just transition towards a climate-resilient future. *Climate and Development, 17*(7), 657–669. https://doi.org/10.1080/17565529.2024.2437133

Harpur, P. (2012). Embracing the new disability rights paradigm: The importance of the convention on the rights of persons with disabilities. *Disability & Society, 27*(1), 1–14. https://doi.org/10.1080/09687599.2012.631794

Heller, T., Schindler, A., Palmer, S. B., Wehmeyer, M. L., Parent, W., Jenson, R., Abery, B.H., Geringer, W., Bacon, A., & O'Hara, D. M. (2011). Self-determination across the life span: Issues and gaps. *Exceptionality, 19*(1), 31–45. https://doi.org/10.1080/09362835.2011.537228

Holburn, S., & Cea, C. D. (2007). Excessive positivism in person-centered planning. *Research and Practice for Persons with Severe Disabilities, 32*(3), 167–172. https://doi.org/10.2511/rpsd.32.3.167

Izutsu, T., Tsutsumi, A., Lu, X., Hashimoto, J., Allotey, P., Dakkak, H., ... & Ito, A. (2019, November). Disability-inclusive disaster risk reduction and humanitarian action: An urgent global imperative. In *United Nations world conference on disaster risk reduction and the progress thereafter*. https://empowerproject.jp/wp-content/uploads/2020/12/2019-WCDRR-Report-and-Thereafter.pdf

Keen, M., Sanderson, D., Osborne, K., Deo, R., Faith, J., & Ride, A. (2022). Area-based approaches and urban recovery in the Pacific: Lessons from Fiji, Solomon Islands and Vanuatu. *Environment & Urbanization, 34*(1), 151–169. https://doi.org/10.1177/09562478211072668

King, J., Edwards, N., Watling, H., & Hair, S. (2019). Barriers to disability-inclusive disaster management in the Solomon Islands: Perspectives of people with disability. *International Journal of Disaster Risk Reduction, 34*, 459–466. https://doi.org/10.1016/j.ijdrr.2018.12.017

Lee, H. C., & Chen, H. (2019). Implementing the Sendai Framework for disaster risk reduction 2015–2030: Disaster governance strategies for persons with disabilities in Taiwan. *International Journal of Disaster Risk Reduction, 41*, 101284. https://doi.org/10.1016/j.ijdrr.2019.101284

Motz, S. (2018). *Article 11: Situations of risk and humanitarian emergencies.* https://doi.org/10.1093/law/9780198810667.003.0012

Mwendwa, T. N., Murangira, A., & Lang, R. (2009). Mainstreaming the rights of persons with disabilities in national development frameworks. *Journal of International Development: The Journal of the Development Studies Association, 21*(5), 662–672. https://doi.org/10.1002/jid.1594

Næss, S., Moum, T., & Eriksen, J. (Eds.). (2011). *Livskvalitet: forskning om det gode liv.* Fagbokforlaget.

Nand, M., Prasad, R., Dwirahmadi, F., Ranse, J., & Mohammadnezhad, M. (2024). *Inclusive Disaster Risk Reduction (IDRR) Efforts in Fiji Islands at the community level: Is it really inclusive?* https://scholarspace.manoa.hawaii.edu/server/api/core/bitstreams/ba40bbe4-7cb0-44e2-92da-67f08a18839e/content

Nations, U. (1994). *Yokohama strategy and plan of action for a safer world.* United Nations: Geneva, Switzerland. https://www.unisdr.org/files/8241_doc6841contenido1.pdf

Njelesani, J., Cleaver, S., Tataryn, M., & Nixon, S. (2012). Using a human rights-based approach to disability in disaster management initiatives. *Natural Disasters,* 21–46. https://doi.org/10.5772/32319

NUMBERS, S. I. (2023). *East Asia and Pacific Region.* East Asia. https://www.unicef.org/appeals/eap/situation-reports

Orsander, M., Mendoza, P., Burgess, M., Arlini, S. M., & Sulaiman, M. (2020). *The hidden impact of COVID-19 on children and families with disabilities.* Save the Children International. https://resourcecentre.savethechildren.net/pdf/the_hidden_impact_of_covid-19_on_children_and_families_with_disabilities.pdf

Pearson, L., & Pelling, M. (2015). The UN Sendai framework for disaster risk reduction 2015–2030: Negotiation process and prospects for science and practice. *Journal of Extreme Events, 2*(01), 1571001. https://doi.org/10.1142/S2345737615710013

Pineda, V. S., & Corburn, J. (2020). Disability, urban health equity, and the coronavirus pandemic: Promoting cities for all. *Journal of Urban Health, 97*(3), 336–341. https://doi.org/10.1007/s11524-020-00437-7

Roy, J., Tscharket, P., Waisman, H., Abdul Halim, S., Antwi-Agyei, P., Dasgupta, P., Hayward, B., Kanninen, M., Liverman, D., Okereke, C. and Pinho, P.F., Riahi, K., & Suarez Rodriguez, A. G. (2018). *Sustainable development, poverty eradication and reducing inequalities.* https://doi.org/10.1017/9781009157940.007

Samaha, A. M. (2007). What good is the social model of disability? *The University of Chicago Law Review, 74*(4), 1251–1308. https://doi.org/10.2307/20141862

Schulze, M. (2010). Understanding the UN convention on the rights of persons with disabilities. *Advocate, 1*, 1–4. http://www.handicap-international.fr/fileadmin/documents/publications/HICRPDManual.pdf

Spurway, K., & Griffiths, T. (2016). Disability-inclusive disaster risk reduction: Vulnerability and resilience discourses, policies and practices. In *Disability in the global south: The critical handbook* (pp. 469–482). Springer International Publishing. https://doi.org/10.1007/978-3-319-42488-0_30

Stjernholm, L. (2024). *Inclusive preparedness: Intellectual disability and disaster risk reduction.* Lunds Universitet/Lunds Tekniska Högskola. https://portal.research.lu.se/en/publications/inclusive-preparedness-intellectual-disability-and-disaster-risk-/

Stough, L. M., & Kang, D. (2015). The Sendai framework for disaster risk reduction and persons with disabilities. *International Journal of Disaster Risk Science, 6*(2), 140–149. https://doi.org/10.1007/s13753-015-0051-8

Strategy, I. *Building disability-inclusive societies in Asia and the Pacific*. https://www.unescap.org/sites/default/files/publications/SDD%20BDIS%20report%20A4%20v14-5-E.pdf

Tozier de La Poterie, A., & Baudoin, M. A. (2015). From Yokohama to Sendai: Approaches to participation in international disaster risk reduction frameworks. *International Journal of Disaster Risk Science, 6*(2), 128–139. https://doi.org/10.1007/s13753-015-0053-6

Twigg, J., Kett, M., & Lovell, E. (2018). *Disability inclusion and disaster risk reduction*. ODI Brief. Note, 1–12. http://dkconf18.modmr.gov.bd/

UNISDR (United Nations International Strategy for Disaster Reduction). (2004). *Review of the Yokohama strategy and plan of action for a safer world*. http://www.unisdr.org/2005/wcdr/intergover/official-doc/L-docs/Yokohama-Strategy-English.pdf

UNISDR (United Nations International Strategy for Disaster Reduction). (2005). *Hyogo framework for action 2005–2015: Building the resilience of nations and communities to disasters*. http://www.unisdr.org/2005/wcdr/intergover/official-doc/L-docs/Hyogo-framework-for-action-english.pdf

UNISDR (United Nations International Strategy for Disaster Reduction). (2013). *UN global survey explains why so many people living with disabilities die in disasters*. http://www.unisdr.org/archive/35032. Accessed 16 Apr 2015.

UNISDR, U. (2012). *Disaster risk reduction and climate change adaptation in the Pacific: an institutional and policy analysis. Suva*, Fiji: UNISDR, UNDP, 76pp. https://www.unisdr.org/files/26725_26725drrandccainthepacificaninstitu.pdf

UNISDR, U. (2015, March). Sendai framework for disaster risk reduction 2015–2030. In *Proceedings of the 3rd United Nations World Conference on DRR, Sendai, Japan* (Vol. 1). https://www.undrr.org/quick/11409

Werner, S. (2012). Individuals with intellectual disabilities: A review of the literature on decision-making since the Convention on the Rights of People with Disabilities (CRPD). *Public Health Reviews, 34*(2), 14. https://doi.org/10.1007/BF03391682

World Bank Group. (2022). *Disability inclusion in disaster risk management: Assessment in the Caribbean region*. https://openknowledge.worldbank.org/server/api/core/bitstreams/fcd13a1b-44d5-5cc0-ba74-a56efc138e8c/content

World Bank. (2018). *The world bank in Pacific Islands*. http://www.worldbank.org/en/country/pacificislands/overview

Chapter 3
Infrastructure, Isolation, and Inaccessibility

Abstract In the Pacific Islands scenario, the first step in this chapter is to analyze the trio of infrastructure, isolation, and inaccessibility in such a way that they lead to a double aspect of vulnerability and resilience in the region. The Pacific Island countries, scattered over huge oceanic distances, are all caught up in the web of geography and infrastructure that makes it impossible to deliver health, education, transportation, and communication services equally—especially to disabled persons. The chapter takes a multidimensional approach: the first one being the geographical isolation aspect that encourages social exclusion; the second is the inadequacy and fragility of infrastructure that magnifies vulnerability; and the last one is the disability inclusion that is still a fringe issue in most resilience frameworks. The author backs up his claim with data gathered from empirical research, regional policy studies, and recent disaster cases such as Cyclones Pam, Winston, and Harold, among others. Besides, subtopics such as geographical isolation and its effects on emergency access, accessibility, and inclusion in built and information environments, and urban versus rural island disparities in Oceania, highlight the complexity of the issue. The chapter argues that the region's resilience can be increased only if the problems of systemic infrastructure and governance inaccessibility are resolved first. It ends by proposing that the Pacific Islands adopt inclusive, community-driven, and context-specific strategies, and that accessibility be viewed as a human right and a prerequisite for sustainable development throughout the region, thereby advocating the reallocation of resources from the latter to the former.

Keywords Disability · Pacific Island Countries · Geographical isolation · Accessibility · Urban-rural disparities

3.1 Introduction

The nations and territories that border the Pacific Ocean are collectively referred to as the Pacific area. The area, commonly called Oceania, has a wide range of civilizations, ethnic groups, levels of economic development, and living conditions. The region's

S. M. Thomas and R. Veerabathiran, *Disability, Disaster, and Resilience in Oceania*, SpringerBriefs in Modern Perspectives on Disability Research,
https://doi.org/10.1007/978-981-95-8535-9_3

health indicators reflect this diversity, with the best performing countries (Australia; life expectancy: 82.8; infant mortality: 4/1000 live births; and New Zealand; life expectancy: 81.6; infant mortality: 6/1000 live births) having life expectancies about 20 years higher and infant mortality up to 15 times lower than the poorest performing countries (Papua New Guinea; life expectancy: 62.9; infant mortality: 57/1000 live births; and Solomon Islands; life expectancy: 69.2; infant mortality: 28/1000 live births) (Horwood et al., 2019).

Pacific Island nations (PICs) are therefore dealing with significant problems resulting from changes in their economies and cultures. Among the elements influencing new lifestyles in the Pacific are migration, monetization, slow and erratic economic growth, the shift from traditional subsistence-based economies to market-based ones, the depletion of natural and social capital, and the decline of traditional social structures. Only two PICs, Niue and the Cook Islands, are on track to meet all of the MDGs by 2015, even though the majority of PICS are middle-income nations with medium-to-high human development and rank among the world's top recipients of aid per capita (UNESCAP et al., 2014).

Geography in Oceania is not just a backdrop; it is a significant factor that creates different risks. There are many island nations in the Pacific, each made up of islands separated by large areas of ocean. The isolation of these areas further increases the time it takes to reach advanced care, makes it even harder to manage evacuations, and sometimes makes EMS (emergency medical services) on land impractical. Research indicates that EMS service areas in Australasia and the Pacific span extensive regions with no access to hospital care within the required time. These regions are rural areas and islands, which are significantly less likely to receive lifesaving interventions during critical windows after injury or acute illness (Lilley et al., 2019). For people with disabilities, these geographical limitations not only affect but also determine their personal mobility, communication, and support requirements. The evacuation routes that are made of narrow footpaths, steep tracks, or boat landings are not open to many wheelchair users and others with mobility impairments. Information about storms about to hit or evacuation orders, when provided only in oral or very centralized formats, may not reach people with hearing or cognitive impairments, or those living alone on remote islands. The operational realities of search and rescue, aeromedical retrieval, and emergency sheltering in archipelagic contexts thus systematically further disadvantage the already marginalized. International and regional reviews are increasingly vocal about the need for context-specific solutions like prepositioned community-based care, equal distribution of point-of-care technologies, and evacuation planning that clearly provides for diverse impairments so that the disparities can be reduced (Aldoukhi et al., 2023; Koste et al., 2023; Lilley et al., 2019).

Infrastructure, including electrical grids, water and sanitation systems, health facilities, transportation networks, and telecommunications, among others, plays a significant role in shaping disaster risk and resilience. To start with, it affects social interaction and estimated access to essential services in everyday situations, which are given to a select few; the problems in this area make it even harder for persons with disabilities to cope with their disadvantages in the fields of work, education, and health. Moreover, it is the infrastructure that determines communities' capacity to bounce back and the speed and fairness of disaster response. One runway in an airport, one hospital on the main island, and one ferry link connecting many far-off islands are just a few examples of critical facilities in the Pacific region that so often contain single points of failure. When those elements are compromised, whole populations might get isolated. According to the Sendai Framework and regional resilience strategies, accessible emergency services and inclusive infrastructure are the core elements to ensuring that disasters do not become catastrophes for persons with disabilities (UNISDR, 2015).

It is not only infrastructure that explains the differences but also political and institutional decisions, budget allocations, planning priorities, and the degree to which disability is included in national disaster laws and policies. The Pacific region has made significant policy advances: regional frameworks such as the Framework for Resilient Development in the Pacific and national commitments under the Sendai Framework highlight the importance of inclusion and the focus on vulnerable groups. However, there are still implementation gaps. Reports and thematic reviews highlight difficulties with data (insufficient disability-disaggregated data for planning), consultation (insufficient involvement of disability organizations), and financing (insufficient funds allocated to making infrastructure and services accessible). Unless these governance deficits are addressed, investments in resilient infrastructure may continue to fail the very people who need them most. The challenges can only be met by integrating technical investments with procedural changes and community-level innovations. There are a few principles that emerge from regional literature and evaluations of post-disaster situations, such as redundancy and decentralization, accessibility by design, community-led planning and participation, and targeted investments to close urban–rural gaps.

The chapter covers the idea development of the connectivity between the three factors, which are infrastructure, isolation, and inaccessibility, in the community's adaptation and vulnerability in Oceania. The chapter elaborates on the discourse by portraying the precariousness of the infrastructure systems brought about by the geographical isolation of the majority of island nations, which manifests as increased disaster risks and slow provision of necessary services, especially for the most affected groups, like disabled persons, by the use of recent evidence, regional experiences, and policy frameworks. In this scenario, the conclusion of the discussion will highlight the urgent need to respond to the resilience-building challenge through an inclusive perspective that recognizes accessibility and its spatial, social, and institutional dimensions. The upcoming conversations aim to clarify these intricate associations, uncover the disparities in the society that lead to the effects of the disasters, and portray the opportunities for creating strategies that are not only resilient but also equitable in the Pacific region regarding accessibility, community involvement, and context sensitivity.

3.2 Geographical Isolation and Its Impact

Geographical isolation is still the main factor that causes inequitable distribution of health, education, and emergency services in the Pacific region. The nations and lands of Oceania face the challenge of being separated by vast oceanic distances, and this means that some communities are even placed on faraway islands that are hundreds of kilometers away from the nearest administrative or medical centers. This situation of geographical fragmentation shows itself in the unequal distribution of infrastructure and service delivery that is considered basic, with facilities such as hospitals, rehabilitation centers, and communication networks being distributed unequally. For people with disabilities, the barriers become even more acute. The physical distance from medical facilities and the like impedes access to timely health interventions, assistive technologies, and rehabilitation services; thereby, the disabled suffer the most, as they face complete restrictions on their participation in community and governance processes (King et al., 2019).

The genesis and progression of mental problems have long been thought to be influenced by geographic isolation. A sizable body of research has proved such a link. However, researchers' ability to rule out competing theories and establish causal relationships has been hampered by study designs. Numerous studies on the involvement of socio-cultural and other environmental factors, especially socioeconomic position, in the genesis and progression of various forms of mental illness have been published in recent years. With few exceptions, the corpus of evidence has not been able to refute alternative theories like spuriousness and reverse causality, despite the occasional hypothesis that geographic isolation contributes to the higher prevalence of mental diseases. Confusion with socioeconomic status may be one example. A variety of helpful treatments might be further developed to promote mental health in remote areas if geographic isolation increases the likelihood of mental impairment. These include telemedicine services, a range of networking, outreach, and self-help techniques, group work, and community-building initiatives like psychiatric clubhouses, and basic amenities like transportation, particularly for those living in remote regions (Holt-Lunstad, 2017; Lubben, 2017; Silva et al., 2016).

Hudson and Doogan's (2019) study made significant contributions to understanding the geographical isolation that remote areas face, as well as the infrastructural inequality closely related to disaster and disability issues in Oceania. The authors present a new and persuasive argument according to which remoteness is a significant factor for higher rates of mental disability, irrespective of SES, social isolation, and income inequality. They analyzed data from 2960 U.S. counties and used both ordinary least squares (OLS) and two-stage least squares (2SLS) regression with instrumental variables (IV) techniques to reveal that isolation is responsible for a considerable amount of the variation in mental disability rates ($\beta = 0.370$, p <0 .001) even after the factors were controlled for. The authors' study draws attention to the isolation factor as a contributor to health differences and to its power by proposing that the mere fact of being far away, through the absence of service access, social support, and resources, can make disability and poor mental health more cosmopolitan (Hudson & Doogan, 2019).

These discoveries have huge implications for the Pacific region. The study suggests that isolation impacts health in the same way as poverty and social deprivation, but it is a singular and measurable health inequality factor that counts. The situation in Oceania is comparable with that of a remote area in the United States, where the physical separation among islands, inadequate transportation, and urban services distributed in one place bring everything to the same rural scenario of the study; thus, the results highlight the problem of reaching medical help for disabled persons. Rural U.S. counties had the least access to regular health services and the highest rates of psychological distress (Arcury et al., 2005; Tittman et al., 2016), while many Pacific Islands communities were already suffering and dealing with logistics and infrastructure problems that not only made them more vulnerable to disasters but also cut off continuity of care. These discoveries are very much significant in the Pacific setting. The scientists highlight through their study that the impact of being cut off from the world is not merely a poverty or social neglect issue but instead an extraordinary and measurable health inequality factor. The situation in one of the most important discoveries of the research, that there is a negative relationship between isolation and both services and community support, is very much in line with the longing for medical and social care that has been so common in most of the Pacific communities, where emergency services and social care networks are dispersed across immense oceanic areas.

In a similar manner, Hudson and Doogan (2019) indicate that the mental health repercussions of isolation can only be dealt with successfully via multi-faceted approaches, which encompass both the development of infrastructure and the involvement of the community. Telehealth expansion, transportation subsidies, and mutual aid networks were singled out as the most essential measures in reducing the impact of distance. In the Pacific, similar methods such as decentralized emergency planning, investment in fare-free and island-hopping transport systems, and culturally tailored community networks for people with disabilities could all contribute to building resilience. The authors' claim that isolation results in "neglect with impunity" (Hudson & Doogan, 2019; Zavaleta et al., 2014) is highly pertinent for the case of Oceania, where the distance factor often leads to the lack of political and infrastructural attention.

McIver et al. (2016) emphasize the fact that the widespread presence of PICs across the globe has made transportation and communication very complicated in terms of logistics, high in cost, and greatly dependent on weather conditions, all these factors having a direct impact on the provision of emergency medical services and the carrying out of disaster response operations.

Besides the logistical challenges, the widely scattered geography of the Pacific has been the root cause of the areas with the most population and economic activity being mainly urban centers and the capital islands. The aforementioned urban bias in development planning has resulted in infrastructural inequities between urban and outer-island populations. Consequently, the gap between the urban and outer island areas in disaster preparedness and recovery capacity becomes very large. The rural and outer-island communities have to endure many times to get through the cycle of getting humanitarian aid during emergencies; furthermore, the disabled persons are usually the ones who face the most obstacles in getting the support that they need. Research conducted by the Secretariat of the Pacific Community (SPC, 2017) and UNESCAP (2014) has revealed that persons with disabilities living in isolated areas are frequently cut off from the benefits of early warning systems, evacuation procedures, and post-disaster rehabilitation programs because of poor communication networks, lack of accessible transport, and no local disability-inclusive emergency planning. The oceanic and air transport that is being relied upon are both weather-sensitive; hence, it is possible that remote areas could be completely isolated from the supplies and medical help for a long time after a disaster, such as the cases of Cyclones Pam (2015) and Winston (2016) (SPC, 2017; UNESCAP, 2014).

In March 2015, Cyclone Pam hit with the full force of a Category 5 tropical cyclone and followed its path through Vanuatu. The cyclone not only destroyed the housing but also left the whole area without any communication for weeks. The destruction of roads and health facilities made it impossible for many outer islands to receive supplies by road or sea for weeks. This circumstance made it even more challenging to supply cuts to the remote communities. Communications infrastructure destruction significantly delayed needs assessments and prolonged targeted assistance to vulnerable groups. Furthermore, the cyclone was so powerful that the transportation of goods to the most isolated communities might mean having to transfer from the larger vessels to the smaller boats or using airlifts logistical steps that are both expensive and slow; moreover, those living in isolated villages with mobility or hearing impairments were at an even higher risk since evacuation planning and relief distributions were not always done inclusively. Thus, the response to Cyclone Pam illustrates how the loss of infrastructure and remoteness together create very severe inaccessibility issues (Hallwright & Handmer, 2019).

The geographical isolation of the Pacific is, a case, not only by the distances but also by the institutional and structural conditions that exhibit how the priorities in infrastructure, governance, and policy are organized. Isolation does not only restrict the movement of people and goods but also creates barriers for the flow of information, policy coordination and social support systems, which are all crucial for inclusion. For instance, in the situation of disabled people, their under-representation in national disaster management activities and the unavailability of adaptive technologies that could mitigate the impact are some of the adverse effects of isolation. Hence, it becomes imperative to devise a two-pronged approach in tackling the inequalities; the first one being to establish inter-island connectivity through the development of transport and communication networks that are easy to access, and the second being doing disability inclusion systematically integrating at all stages of planning from preparedness to recovery. Breaking the isolation is the key to attaining resilience and human security in the Pacific, especially as climate change is expected to worsen both environmental and social risks. UNESCAP (2014) and McIver et al. (2016) point out that the first step towards achieving resilience and security is breaking the isolation, which would make it easier for people to at least cope with climate change impacts. Besides, making geographic isolation a source of localized resilience and empowerment rather than a hurdle would be through the region-specific strategies of setting up decentralized emergency healthcare systems, community-based rehabilitation programs, and providing inclusive communication technology, among others.

Let us consider the Solomon Islands as a case study, among the nations most affected by geographical isolation and poor infrastructure, and thus one of the worst sufferers of the combination of the two. The vast distances in the island nations make transportation, communication, and service delivery very difficult. The outer islands can be reached only by irregular freight or air transport, which is sometimes interrupted by bad weather, resource shortages, or both. The health, education, and extinction (World Bank, 2019) systems that service the differently-abled, who are already hampered by mobility and communication barriers, are the most affected by the situation. Honiara, the capital city, is the center of infrastructure development, while rural and outer-island communities are coping with poor electricity supply, poor roads, and a lack of health and rehabilitation facilities. It is for this reason that disabled people in these areas are unlikely to be included in emergency planning and disaster support (ADB, 2020; UNESCAP, 2014).

During natural disasters such as Cyclone Harold in 2020 or the 2014 flash floods, remote areas were utterly cut off from relief operations. The airstrips were destroyed, bridges were submerged, and there was no communication. These infrastructural frailties would not only delay but also prevent the delivery of emergency services, for which resources such as assistive devices, medications, and disability-specific rehabilitation support were needed (SPC, 2017). The existing situation is made worse by the fact that the policies do not adequately reflect the disabilities; they do not recognize disabilities as a valid category, and hence, there is no one representing the disabled community in the decision-making process (McIver et al., 2016). This situation, therefore, indicates a profound structural inequity. The factors contributing to this isolation are not only geographical but also institutional, which has led to the continuous exclusion of persons with disabilities from the national planning and development frameworks.

Focusing on the power given to local communities in the form of resilient strategies that are inclusive and decentralized, the researchers along with the regional authorities attempt to face the linked difficulties. In addition, inter-island connections will have to depend on transport infrastructures and communication systems that are not only resilient and responsive but also dependable. Disability planning that is incorporated in community-based disaster preparedness and early warning systems can significantly enhance the adaptive capacity of remote provinces. The Framework for Resilient Development in the Pacific (SPC, 2017) indicates that the resilience of the Solomon Islands will not be fair unless the investments are made not only in the physical infrastructure but also in the access of the institutions that guarantee people with disabilities to be seen, included, and assisted in each step of the resilience-building process (Gartrell et al., 2018).

3.3 Accessibility and Inclusion in Built and Information Environments

The physical environment, transport facilities, knowledge, and access to communication systems are the pillars of a society that is inclusive and equitable for all. Accessibility is the basic right for persons with disabilities to assert their rights and interact with the community. The Pacific area is experiencing even more difficult times in terms of access because of its scattered islands, lack of good transport connections, and short funds. The United Nations Economic and Social Commission for Asia and the Pacific (UNESCAP) claims that accessibility should not be considered a mere technical measure anymore but rather a precondition for participation, self-reliance, and inclusion (UNESCAP, 2019).

People with disabilities demand accessibility as a right, not just a luxury. The rights to education, to get a job, to vote, and to receive relief during disasters are not possible without it. Despite the undertakings, inaccessible spaces continue to exist and fully restrict people's participation in social and economic life. In the Pacific region, the situation is worse due to the remoteness of the places and the lack of good infrastructure. The principle of universal design, which promotes the creation of environments accessible to everyone regardless of ability, is a perfect guide for solving these problems. The accessibility planned in urban, rural, and even remote areas not only helps the disabled but also the whole community in terms of safety, mobility, and usability improvements. Access audits, systematic evaluations of built environments, are crucial to ensuring that accessibility regulations are observed throughout the planning, design, construction, and even maintenance phases of a project (ESCAP, 2017).

Accessibility investment must cover all areas of the environment and the information ecosystem so that everyone, regardless of physical or sensory limitations, can enjoy the same interaction quality across both spaces and systems. The Incheon Strategy (2013–2022) and the Beijing Declaration and Action Plan promote accessibility by recommending the use of universal design principles across all types of planning for both urban and rural areas. The Incheon Strategy (2013–2022) has recently concluded regional accessibility assessments in the Pacific and Oceania, revealing widely different levels of accessibility. It is stated that around 66.5% of government buildings and 70.6% of international airports in the region are accessible, yet there are considerable disparities between countries, ranging from 0 to 100% in accessibility. Although accessibility assessments cover a majority of cases for the mobility-impaired only, they do not cater to people with sensory, intellectual, or psychosocial disabilities. A few Pacific Island nations, like Vanuatu, Micronesia, and Tonga, have had their accessibility conditions audited, which showed that public buildings and transport infrastructure have a long way to go before meeting international standards. Access to airports, primarily used for evacuation during a disaster and inter-island travel, is one of the main factors that define a region's resilience. Making these transit points accessible will help build more inclusive emergency and mobility networks across Oceania. Whereas, in the case of small island nations like Fiji and Solomon

Islands, the problems of poor infrastructure, such as uneven pavements and the failure of communication systems, plus the non-existence of mobility support, have created barriers that significantly reduce the movement of people with disabilities, especially in the case of emergencies (ESCAP, 2017; Strategy; UNESCAP, 2019).

Beyond looking at individual incidents, studies over the Pacific indicate that medical centers and primary care hospitals are often built with little redundancy and are prone to extreme weather. If the health infrastructure is damaged, routine chronic-care treatment (which includes disability treatment) will be stopped, and emergency-care capacity will be reduced. Regional assessments indicate that a single damaged unit can leave an entire province without proper healthcare, resulting in a series of public health problems. Building up health infrastructure and proper planning are thus of utmost importance for both regular access and disaster resilience (Natuzzi & Zealand, 2024).

Damage to essential transportation links, bridges, auxiliary roads, ports, and airstrips has consistently hindered relief efforts in the area. The case studies indicate that communities without access to an all-year road or that could only access seasonal boat service experienced prolonged isolation after the storms, with consequent food insecurity, interruptions in medication supply, and delayed evacuation of injured people. Such transportation collapses hit small-island and rural inland communities, where affluence and social services are already limited, thereby increasing the impact of cut lifelines (Indrawansa, 2015).

Apart from the physical infrastructure, the availability of information and communication technologies (ICTs) is also a crucial factor in promoting and enabling participation and inclusion. Despite the increase in internet usage and mobile connections, the digital world remains out of reach for many disabled people because the web is not designed to be accessible, screen readers are not compatible, and there is no captioning for media. UNESCAP (2019) notes that less than 50% of public websites in the Pacific region are compliant with the minimum accessibility standards set by the Web Content Accessibility Guidelines (WCAG 2.0). Inaccessible information has severe consequences during disasters, as inaccessible alerts and public announcements can leave people unaware and put them at risk. Therefore, the digital divide remains a significant challenge, with several Pacific Island states reporting extremely low accessibility to public websites, media, and educational resources (Mines et al., 2020; UNESCAP, 2019).

A mere 40% of public websites in the area are thought to have the required accessibility standard, while the corresponding figure for television news programs' accessibility features is only 41%, with features such as captioning or sign language interpretation being provided. Such accessibility issues result in the delay of information delivery to people with disabilities, which, in turn, causes them not to be able to receive the information on time, especially during disasters or public health emergencies. Moreover, many people need assistive technologies but are unable to access them; more than one-third of the disabled people in the countries reporting do not have access to the devices that would allow them to live independently and participate in daily activities. For Pacific Island countries, it is a necessity to supply assistive devices, facilitate transport for all, and create a digital infrastructure that is inclusive, to close the divide between isolation and participation, and between vulnerability and resilience; the use of accessible information and communication technologies (ICTs) like text-to-speech systems, captioning, and sign language interpretation will be a prudent and economical strategy in promoting inclusion and enhancing disaster resilience in the region's widely scattered island territories (UNESCAP, 2016; UNESCO, 2013).

A very critical factor is also the application of reasonable accommodation measures that go along with accessibility and universal design. These measures enable people with disabilities to access education, employment, and healthcare, even where full accessibility has not yet been established. For instance, the use of sign language interpreters, accessible transport services, or assistive devices can temporarily close gaps while larger accessibility projects are underway. In addition, the active involvement of persons with disabilities in the planning and monitoring of accessibility projects is a must. The CRPD (2006) and General Comment No. 2 (2014) underscored that accessibility is not a matter of charity but a legal requirement, and that persons with disabilities should be regarded as partners and decision-makers at every developmental stage (Persson et al., 2015).

The Pacific region gaining easier access is a matter not only of ethics but also of profit. The construction of skyscrapers, the building of roads, and the installation of information systems will open up new social and economic advantages through improved mobility, participation, and the whole society's resilience. UNESCAP (2019) states that taxpayers' funds should not be used to maintain disability caused by inaccessible services or facilities. Instead, the government and the regional institutions should take on the responsibility to determine the accessibility bar that should be gone, and do so very slowly, in every area, transportation, communication, housing, education, and so forth. In addition, national policies must conform to international treaties such as the CRPD and the 2030 Agenda for Sustainable Development. The entire evolution of accessibility in Pacific societies will ensure that no one is excluded, especially in times of disasters, when only inclusive infrastructure and information systems may actually be lifesaving. Figure 3.1 depicts the comparison of the various existing barriers and the required improvements in the context of disability inclusion.

3.4 Urban vs Rural Island Disparities in Oceania

The Pacific region has a complex geography of varying levels of development, where the difference between cities and rural or isolated island communities is not only a matter of physical location but also of the factors involved. Living in rural areas or on remote islands corresponds to a stark difference between these places and urban centers, which are very much alive and have even more to offer in terms of activities like job opportunities in health and education, as well as better governance. This urbanization trend is noticeable in some PICs, especially in the capitals of Fiji, Papua New Guinea, Samoa, and the Solomon Islands. Besides, the population and urban

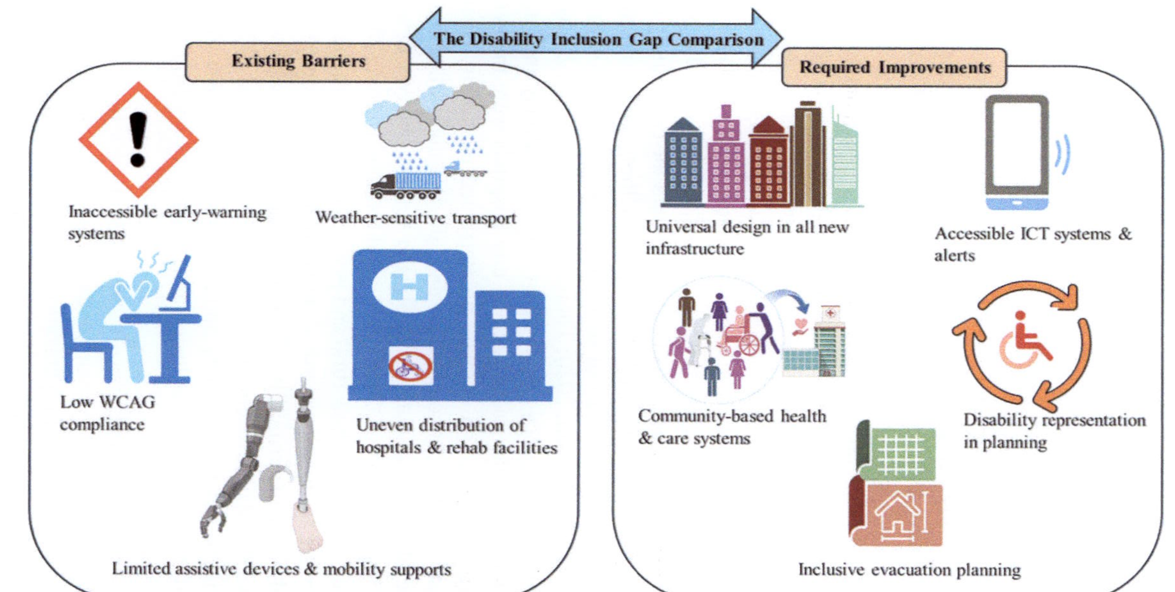

Fig. 3.1 Comparison of existing disability-related barriers and the improvements required to achieve inclusive disaster management systems. This figure compares the existing barriers that persons with disabilities face during disaster situations with the improvements needed to close the disability inclusion gap. The left panel illustrates current challenges, including inaccessible risk information, environmental hazards, stress, and exclusion in information access, non-inclusive healthcare facilities, and the loss or inaccessibility of assistive devices. The right panel highlights necessary improvements such as accessible urban infrastructure, inclusive communication technologies, community participation in preparedness, disability-inclusive healthcare systems, and the integration of universal design principles in rebuilding and planning. This comparison emphasizes the need for systemic changes to create resilient and inclusive disaster management frameworks

facilities are still very concentrated in these cities and as a result, there is a significant difference between the urban and rural areas in terms of urban infrastructure development. People with disabilities are becoming the most affected by the situation since they are less able to access the services and are not given assistive devices and opportunities to interact with others. In the majority of the rural and remote island areas, the transport systems are still very underdeveloped, roads are either unpaved or become impassable during heavy rains, and communication between the islands is maintained through ships whose schedules are very irregular. All these factors contribute to worsening the already existing problem of social and physical isolation (Campbell, 2019; Connell, 2013; Smith, 2023).

In the cities, the situation is still not perfect when it comes to access for people with disabilities, but at least it is slowly being recognized as part of society, with policy reforms and other interventions paving the way. The implementation of national disability-inclusive development strategies in nations like Fiji and Samoa is a step forward that has enabled people with disabilities to have better access to education, vocational training, and social welfare services. Nevertheless, the expansion of these advancements is mostly happening in capital cities. To illustrate, Suva is home to rehabilitation centers, accessible schools, and organizations advocating for people with disabilities. At the same time, rural areas in Vanua Levu and the outer islands of Lau are often neglected and lack even basic medical services, let alone specialized services for people with disabilities. Hence, it is the families and the community of the disabled people living in the areas that provide the support, although it is cultural, it may not be enough to replace the professional services and adaptive technologies that are missing (Pacific Disability Forum, 2020; Thomas, 2011; Vira et al., 2025).

Infrastructures for health and emergency responses were still very slowly unifying in urban and rural areas. Generally, the Pacific capital city hospitals were the best equipped, staffed with professionals, and enjoyed uninterrupted electricity, clean water, and telecommunication access. In contrast, medical devices and skilled staff were still missing in several rural and outer-island health centers (McIver et al., 2016). The instances of Cyclone Winston (2016) in Fiji and Cyclone Harold (2020) in the Solomon Islands are examples of how rural and isolated regions were the last to receive aid, due to damage to airstrips, washed-away roads, and disrupted sea transport. These logistical problems primarily affected disabled people, who were considered the last to be evacuated or to receive help (SPC, 2017; Veitata et al., 2021). This clearly shows how disparities in geography and infrastructure increase social vulnerability, thereby reinforcing cycles of exclusion and dependency (Wilson et al., 2023).

The division between urban and rural areas is evident in access to education and employment opportunities. Usually, urban centers provide disabled children with special education, non-governmental organizations, and various facilities, e.g., ramps or accessible toilets, as they already have specially trained teachers, learning materials, etc., that are lacking in rural schools (UNESCAP, 2019). Hence, the children with disabilities in the countryside and on the islands have a lower chance of attending and completing school, and this in turn limits their future chances of getting a job. The availability of jobs for people with disabilities is also limited in urban economies dominated by service and administrative sectors, while rural areas still rely on subsistence agriculture and informal work, which lack accessibility adaptations (Taneja-Johansson et al., 2023).

Apart from access, information, and ICTs have become significant factors in the Pacific's inclusion and resilience. UNDP (2021) stated that digital connectivity remains highly unequal, despite a significant increase in both mobile and internet penetration rates, and that rural and outer-island regions suffer the most from slow connections, high prices, and frequent outages. The digital divide acts as a barrier that restricts access to information, online teaching, and telemedicine, and on top of that, people with disabilities are the ones who find it most difficult to obtain the services they need. ICTs, such as text-to-speech systems, captioning, and assistive software, are often unavailable in rural areas, thereby deepening the information gap. UNESCAP (2019) stated that the provision of equal digital access is a requirement not only for social inclusion but also for effective communication and dissemination of early warning messages during disasters in the unapproachable island territories.

The divergences observed have been, to some extent, shaped by the policy and governance climate in Oceania. Centralized governance, which is usually carried out in the capitals, creates a gap between policy and remote provinces, where actual conditions differ. The laws and regulations that secure the rights of disabled people are mostly confined to certain areas, limiting their influence in the countryside. Most rural local governments lack both the necessary expertise and the funds to implement and make the necessary infrastructural changes, as well as to provide inclusive services. Additionally, there is a lack of disaggregated data by area and disability status, which prevents the proper planning and monitoring of the inclusion projects being carried out. Therefore, promoting decentralization in governance and ensuring local participation in decision-making are necessary to achieve balanced development and inclusive resilience (Cattaneo et al., 2022; Pacific Islands Forum Secretariat, 2020; Van Assche et al., 2020).

However, rural and island communities still have their strengths, which can, by and large, promote resilience. As a rule, people living in remote areas and on islands have strong interpersonal relationships, strong family and friend support, and indigenous knowledge, which in turn makes them less dependent on formal systems. Furthermore, community-driven actions such as disaster preparedness training sessions for all and peer-support groups have been seen to contribute to increased local adaptive capacity. Applying these community-based methods alongside national disability policies will go a long way towards closing the gap between urban and rural areas. A case in point is the work in Vanuatu and Samoa, where an alliance of local disability organizations and disaster management authorities has led to improved access to shelters and to evacuation planning (Bohensky et al., 2024; Latai-Niusulu et al., 2024; Villa, 2021).

The Pacific Islands' governments, along with regional organizations, should first solve the problem of inequality between urban and rural areas and then make the investment in inclusive infrastructure the first priority. Additionally, the area, where the programs are further developed, must comprise the distribution of assistive devices, inclusive education, and internet access for the rural and outer-island areas as part of the series of activities. The Framework for Resilient Development in the Pacific (FRDP, 2017–2030) and the Pacific Framework on the Rights of Persons with Disabilities (2016–2025) are two regional initiatives that highlight the necessity of bringing remotely situated populations into the development cycle. The realization of these frameworks will depend a great deal on the allocation of resources, political backing, and the involvement of different groups of persons with disabilities in the process (Kiddle et al., 2017).

Thereby, reducing the urban–rural and island gaps in Oceania is fundamental to achieving the 2030 Agenda for Sustainable Development and the Convention on the Rights of Persons with Disabilities (CRPD, 2006). Providing infrastructure, healthcare, education, and digital technology to all, without discrimination, can change the situation and make people more powerful and less defenseless. Hence, the Pacific inclusive development has to go the way of not only the cities but also the entire region with its different needs, making it acceptable for everyone, moving the bottlenecks of access to opportunities, and creating ways that cater to all the people, no matter where they reside or what problems they face.

3.5 Conclusion

Pacific Island communities live their lives through infrastructure, isolation, and inaccessibility, which are the primary factors that determine their ability to endure and recover from crises. The distribution of these islands over a large area accentuates disparities in the provision of services and governance, thus leaving populations living on outer islands and in rural areas, including persons with disabilities, at the bottom of the list for policy enactment and emergency management. The lack of accessible infrastructure and communication systems not only restricts people's involvement in the decision-making processes but also keeps on reinforcing the patterns of exclusion and dependency that have been created before. The situations in the Solomon Islands and Vanuatu illustrate that, for example, the humanitarian aid after the disasters often does not get to the isolated locations quickly enough, which shows an important relationship between geography and resilience.

Additionally, accessibility issues should not only be considered in the context of the removal of physical barriers but mainly through the lens of inclusive information systems, digital connectivity, and equitable participation in governance. The line separating isolation from inclusiveness can be crossed with the help of the development and upkeep of universally designed settings as well as the granting of assistive technologies. The empowerment of local communities through decentralized planning and capacity-building can contribute to the adaptation of responses to environmental and social vulnerabilities even more. The Regional frameworks like the Framework for Resilient Development in the Pacific (FRDP, 2017–2030) and the Pacific Framework on the Rights of Persons with Disabilities (2016–2025) are some of the policy foundations that already exist. However, their success will still rely heavily on strong political will, sufficient funding, and the ongoing participation of disability organizations at all levels.

The gap between urban and rural areas needs to be addressed without delay. On the one hand, Pacific cities are becoming increasingly progressive in terms of access and service provision, whereas on the other hand, the remote islands are still struggling to obtain the basic resources they need. The only way to balance this is through infrastructure investments specifically for this purpose, the development of transportation networks that are easy to use, and the provision of digital access that is equitable enough to turn exclusion into empowerment. One way to extend ICTs (e.g., captioned communication, sign language interpretation, and text-to-speech systems) is to provide participation and disaster preparedness across remote areas as a cost-effective measure. The resilience of Oceania, in the end, will entirely rely on the involvement of the various communities. If access is considered the main factor in sustainable development, then everyone would receive assistance during disaster management phases, and no one would be left out, especially disabled individuals. The transition towards inclusive infrastructure, universal design, and participatory governance will lower the disaster risk and, at the same time, create stronger, better-connected, and more independent communities in the Pacific. The insights from this chapter highlight the need, in terms of rights, society, and practicality, to incorporate accessibility and disability inclusion at all levels of regional development and disaster resilience policy.

References

Aldoukhi, M., Angel, M., Bare, L., Blacher, D., Cawi, J., Chabanel, T., Ciccone, M., Clapper, K., El Jai, S., Ferrer, A. and Figueroa, P., & Mesa, J. C. (2023). *Access of persons with disabilities to public ground transportation and roadways*. https://docs.lib.purdue.edu/cgi/viewcontent.cgi?article=1005&context=ugcw

Arcury, T. A., Gesler, W. M., Preisser, J. S., Sherman, J., Spencer, J., & Perin, J. (2005). The effects of geography and spatial behavior on health care utilization among the residents of a rural region. *Health Services Research, 40*(1), 135–156. https://doi.org/10.1111/j.1475-6773.2005.00346.x

Asian Development Bank (ADB). (2020). *Solomon Islands: Country partnership strategy 2021–2025*. Manila:ADB. https://www.theprif.org/sites/theprif.org/files/documents/ADB%20Pacific-approach-2021-2025.pdf

Bohensky, E. L., Butler, J. R., Bedford, K., Rainbird, J., McGrath, V., Busilacchi, S., Skewes, T. D., Maru, Y. T., Hunter, C., Schoon, M., Nai, T. F., & Mosby, H. (2024). "Going back to what really held us together": Re-adaptation as resilience in the Torres Strait Islands, Australia. *Ecology and Society, 29*(4). https://doi.org/10.5751/ES-15562-290442

Campbell, J. R. (2019). Climate change and urbanization in Pacific Island Countries. *Policy Brief*, (49).

Cattaneo, A., Adukia, A., Brown, D. L., Christiaensen, L., Evans, D. K., Haakenstad, A., McMenomy, T., Partridge, M., Vaz, S., & Weiss, D. J. (2022). Economic and social development along the urban–rural continuum: New opportunities to inform policy. *World Development, 157*, 105941. https://doi.org/10.1596/1813-9450-9756

Connell, J. (2013). *Islands at risk?: Environments, economies and contemporary change*. Edward Elgar Publishing. https://doi.org/10.4337/9781781003510

CRPD. (2006) Convention on the Rights of Persons with Disabilities and Optional Protocol. https://www.un.org/disabilities/documents/convention/convoptprot-e.pdf

ESCAP, U. (2015). *Disability at a glance 2015: Strengthening employment prospects for persons with disabilities in Asia and the Pacific.* https://medbox.org/document/disability-at-a-glance-2015-strengthening-employment-prospects-for-disability-at-a-glance-2015-persons-with-disabilities-in-asia-and-the-pacific

ESCAP, U. (2017). *Building disability-inclusive societies in Asia and the Pacific: Assessing progress of the Incheon strategy.* https://www.unescap.org/publications/building-disability%E2%80%91inclusive-societies-asia-and-pacific-assessing-progress-incheon

ESCAP, U. (2019). *Disability at a Glance 2019: Investing in accessibility in Asia and the Pacific: Strategic approaches to achieving disability-inclusive sustainable development.* https://www.unescap.org/publications/disability-glance-2019

Gartrell, A., Jennaway, M., Manderson, L., Fangalasuu, J., & Dolaiano, S. (2018). Social determinants of disability-based disadvantage in Solomon islands. *Health Promotion International, 33*(2), 250–260. https://doi.org/10.1093/heapro/daw071

Hallwright, J., & Handmer, J. (2019). Accountability and transparency in disaster aid: Cyclone Pam in Vanuatu. *International Journal of Disaster Risk Reduction, 36*, Article 101104. https://doi.org/10.1016/j.ijdrr.2019.101104

Holt-Lunstad, J. (2017). The potential public health relevance of social isolation and loneliness: Prevalence, epidemiology, and risk factors. *Public Policy & Aging Report, 27*(4), 127–130. https://doi.org/10.1093/ppar/prx030

Horwood, P. F., Tarantola, A., Goarant, C., Matsui, M., Klement, E., & Umezaki, M. (2019). Health challenges of the Pacific region: Insights from history, geography, social determinants, genetics, and the microbiome. *Frontiers in Immunology, 2019*(10), 2184. https://doi.org/10.3389/fimmu.2019.02184

Hudson, C. G., & Doogan, N. J. (2019). The impact of geographic isolation on mental disability in the United States. *SSM-Population Health, 8*, Article 100437. https://doi.org/10.1016/j.ssmph.2019.100437

Indrawansa, P. P. G. P. P. (2015). *Vanuatu: Vanuatu Cyclone Pam disaster response project.* https://coilink.org/20.500.12592/m0g0qp

Kiddle, G. L., McEvoy, D., Mitchell, D., Jones, P., & Mecartney, S. (2017). Unpacking the Pacific urban agenda: Resilience challenges and opportunities. *Sustainability, 9*(10), 1878. https://doi.org/10.3390/su9101878

King, J., Edwards, N., Watling, H., & Hair, S. (2019). Barriers to disability-inclusive disaster management in the Solomon Islands: Perspectives of people with disability. *International Journal of Disaster Risk Reduction, 34*, 459–466. https://doi.org/10.1016/j.ijdrr.2018.12.017

Kost, G. J., Füzéry, A. K., Caratao, L. K. R., Tinsay, S., Zadran, A., & Ybañez, A. P. (2023). Using geographic rescue time contours, point-of-care strategies, and spatial care paths to prepare island communities for global warming, rising oceans, and weather disasters. *International Journal of Health Geographics, 22*(1), 38. https://doi.org/10.1186/s12942-023-00359-y

Latai-Niusulu, A., Taua'a, S., Lelaulu, T., Zari, M. P., & Bloomfield, S. (2024). Working with nature, working with Indigenous knowledge: Community priorities for climate adaptation in Samoa. *Nature-Based Solutions, 6*, Article 100144. https://doi.org/10.1016/j.nbsj.2024.100144

Lilley, R., de Graaf, B., Kool, B., Davie, G., Reid, P., Dicker, B., Civil, I., Ameratunga, S., & Branas, C. (2019). Geographical and population disparities in timely access to prehospital and advanced level emergency care in New Zealand: A cross-sectional study. *BMJ Open, 9*(7), e026026. https://doi.org/10.1136/bmjopen-2018-026026

Lubben, J. (2017). Addressing social isolation as a potent killer! *Public Policy & Aging Report, 27*(4), 136–138. https://doi.org/10.1093/ppar/prx026

McIver, L., Kim, R., Woodward, A., Hales, S., Spickett, J., Katscherian, D., Hashizume, M., Honda, Y., Kim, H., Iddings, S. and Naicker, J., McMichael, A. J., & Ebi, K. L. (2016). Health impacts of climate change in Pacific Island countries: a regional assessment of vulnerabilities and adaptation priorities. *Environmental Health Perspectives, 124*(11), 1707–1714. https://doi.org/10.1289/ehp.1509756

Mines, K., Brentnall, L., Kuambu, A., Kwaram, A., Layton, N., Macanawai, S.,Parasyn, C., Taloafiri, E., Utamapu, F. M., & Mines, K. (2020). *Building sustainable and effective assistive technology provision in partnership: Lessons from the Pacific.* Global perspectives on assistive technology: proceedings of the GReAT Consultation 2019, World Health Organization, Geneva, Switzerland, 22–23 August 2019. Volume 2, 208.

Natuzzi, E., & Zealand, A. N. (2024). *Vulnerability of Pacific Island country hospitals: Critical infrastructure that must be addressed.* CANZPS Oceanic Currents. Available: https://canzps.georgetown.edu/2023/01/04/vulnerability-of-pacific-island-country-hospitals-critical-infrastructure-that-must-be-addressed/ [Accessed 4 Jan 2023].

Pacific Disability Forum (PDF). (2020). *Pacific disability inclusive development report 2020.* Suva, Fiji.

Pacific Framework for Resilient Development. (2016). *Framework for resilient development in the Pacific.* https://pacificdisability.org/wp-content/uploads/2024/04/FRDP_2016_Resilient_Dev_pacific.pdf

Persson, H., Åhman, H., Yngling, A. A., & Gulliksen, J. (2015). Universal design, inclusive design, accessible design, design for all: Different concepts—one goal? On the concept of accessibility—historical, methodological and philosophical aspects. *Universal Access in the Information Society, 14*(4), 505–526. https://doi.org/10.1007/s10209-014-0358-z

Silva, M., Loureiro, A., & Cardoso, G. (2016). Social determinants of mental health: A review of the evidence. *The European Journal of Psychiatry, 30*(4), 259–292. https://www.researchgate.net/publication/312937505_Social_determinants_of_mental_health_A_review_of_the_evidence

Smith, F. (2023). *Early identification and intervention for children with disability in Fiji–current practices and opportunities.* https://mspgh.unimelb.edu.au/__data/assets/pdf_file/0012/4667178/Fiji-ECI-Final-report-March-2023.pdf

SPC, S., & PIFS, U. (2017). UNISDR, & USP.(2016). *Framework for resilient development in the Pacific: An integrated approach to address climate change and disaster risk management (FRDP) 2017, 2030*(4). https://www.hawaii.edu/climate-data-portal/wp-content/uploads/2022/09/FRDP_2016_finalResilient_Dev_pacific.pdf

Strategy, I. *Building disability-inclusive societies in Asia and the Pacific.* https://www.unescap.org/sites/default/files/publications/SDD%20BDIS%20report%20A4%20v14-5-E.pdf

Taneja-Johansson, S., Singal, N., & Samson, M. (2023). Education of children with disabilities in rural Indian government schools: A long road to inclusion. *International Journal of Disability, Development and Education, 70*(5), 735–750. https://doi.org/10.1080/1034912X.2021.1917525

Thomas, P. (2011). Implementing disability-inclusive development in the Pacific and Asia. *Development Bulletin (Canberra), 74.* https://scispace.com/pdf/implementing-disability-inclusive-development-in-the-pacific-ogz0zry78a.pdf

Tittman, S. M., Harteau, C., & Beyer, K. M. (2016). The effects of geographic isolation and social support on the health of Wisconsin women. *Wmj, 115*(2), 65–69. https://www.researchgate.net/publication/301631863_The_effects_of_geographic_isolation_and_social_support_on_the_health_of_Wisconsin_women

UNDP. (2021). Asia and the Pacific. https://www.undp.org/asia-pacific

UNESCAP, et al. (2014) *The state of human development in the Pacific: A report on vulnerability and exclusion in a time of rapid change.* https://hdl.handle.net/20.500.12870/3101

UNESCAP. (2019). Asia-Pacific Forum on Sustainable Development 2019. https://www.unescap.org/events/asia-pacific-forum-sustainable-development-2019

UNESCO. (2013). *Global report: Opening new avenues for empowerment—ICTs to access information and knowledge for persons with disabilities.* Paris: UNESCO. https://igualdad.cepal.org/en/digital-library/unesco-global-report-opening-new-avenues-empowerment-icts-access-information-and-0

UNISDR, U. (2015, March). Sendai framework for disaster risk reduction 2015–2030. In *Proceedings of the 3rd United Nations World Conference on DRR, Sendai, Japan (Vol. 1)*. https://www.undrr.org/publication/sendai-framework-disaster-risk-reduction-2015-2030

United Nations Economic and Social Commission for Asia and the Pacific. (2016). *Disability at a glance 2015: Strengthening employment prospects for persons with disabilities in Asia and the Pacific. United Nations*. https://www.unescap.org/sites/default/files/publications/SDD%20Disability%20Glance%202015_Final_0.pdf

Van Assche, K., Hornidge, A. K., Schlüter, A., & Vaidianu, N. (2020). Governance and the coastal condition: Towards new modes of observation, adaptation and integration. *Marine Policy, 112*, Article 103413. https://doi.org/10.1016/j.marpol.2019.01.002

Veitata, S., Miyaji, M., Fujieda, A., & Kobayashi, H. (2021). *Social capital in community response after Cyclone Winston: Case study of three different communities in Fiji*. https://doi.org/10.17608/k6.auckland.13578272.v2

Villa, V. (2021). *Gender equality, disability and social inclusion (GEDSI) considerations for the climate information services for resilient development in Vanuatu (VAN-KIRAP) project*. https://www.sprep.org/sites/default/files/vankirap/Van%20KIRAP_Gender%20Assessment_Final%20Report.pdf

Vira, A. E., Page, A., & Ledger, S. (2025). Lessons from the Pacific: A scoping review of Vanuatu disabilities in education practices. *International Journal of Inclusive Education,* 1–17. https://doi.org/10.1080/13603116.2025.2530633

Wilson, D., Rokoduru, A., Gade, W., & Tawake, K. (2023). Country case study: Fiji containing, mitigating, and responding to COVID-19: Knowledge generation and exchange, preparedness, and response (March 2020 to June 2022). Country CASE STUDy: Fiji containing, mitigating, and responding to COVID-19: Knowledge generation and exchange, preparedness, and response (March 2020 to June 2022), 1–110. https://doi.org/10.1596/40752

World Bank. (2018). *Solomon Islands country diagnostic: Systematic country assessment*. World Bank. https://doi.org/10.1596/29881

Zavaleta, D., & Samuel, K. (2014). *Social isolation: A conceptual and measurement proposal*. https://hdl.handle.net/20.500.12413/11811

Chapter 4
Voices from the Islands: Oral Histories of Disabled Survivors

Abstract This chapter is about the lives of disabled people in the Pacific Island Countries (PICs) who are facing the climate change-induced disasters, which are getting more and more frequent. This chapter on the intersection of scientific data about climate risks and personal narratives is the first to assert that disabled people are suffering because of climate change and poor sociocultural conditions. It highlights the fact that regional surveys and oral testimonies from Samoa, Fiji, Vanuatu, and the Solomon Islands have uncovered the systemic, institutional, and historical reasons that not only limit disaster experience accessibility, poor communication systems, but also the exclusion from preparedness planning and the ongoing colonial and ableist legacies. Moreover, the chapter shows the survival and adaptation techniques relating to indigenous knowledge, community support, inclusive communication, and capability-based justice frameworks. The chapter that puts disability inclusion at resilience's very heart asserts that the Pacific should not only listen to but also involve the disabled persons' voices, leadership, and local knowledge in the sustainable disaster risk reduction (DRR) measures' implementation. In the end, the Pacific Islands must come up with ways of dealing with disasters that are culturally-rooted, society-consistent, and at the same time compatible with global protocols like the Sendai Framework and the Paris Agreement.

Keywords Pacific Island Countries · Natural disasters · People with disabilities · Indigenous knowledge · Resilience

4.1 Introduction

Floods, earthquakes, cyclones, and droughts are ranked among the natural calamities that inflict the most damage and have the widest tolerable human contact, thereby causing economic losses, taking lives, and halting development. The World Risk Report 2012 suggested that the impact of a natural disaster in a country and its severity are mainly determined by the social, economic, and institutional factors in the respective country. Man-made practices such as deforestation, soil erosion, and

S. M. Thomas and R. Veerabathiran, *Disability, Disaster, and Resilience in Oceania*, SpringerBriefs in Modern Perspectives on Disability Research,
https://doi.org/10.1007/978-981-95-8535-9_4

coastal ecosystem destruction have removed the natural barriers and, consequently, have greatly increased the number of disasters. Over and above that, the global warming due to humans has so elevated the natural vulnerabilities that their impacts cannot be lessened at all. Thus, disaster risk reduction is recognized as a problem not only for the humanitarian side but also for the moral and economic side. It requires the execution of preparedness, prevention, and resilience practices, along with more extensive and active policies for sustainable development (World Risk Report, 2012).

There is evidence that more extreme weather events, including heat waves, precipitation extremes, and the potential for more severe storms, are being recorded as a result of global warming (Coumou & Rahmstorf, 2012). By the end of the twenty-first century, the Pacific's predicted frequency of tropical cyclones may follow worldwide trends of fewer tropical cyclones and a probable rise in the proportion of severe storms. Pacific Islanders are accustomed to dealing with adverse weather and climatic conditions. For example, severe drought and hunger in certain Pacific islands, including significant agricultural losses in Fiji and widespread starvation in parts of Papua New Guinea, were attributed to the 1997–1998 El Niño event. Tropical cyclones, increased precipitation, and flooding, along with cyclone-induced surges, have caused huge financial and infrastructural losses, especially in Samoa, the Cook Islands, and Fiji. Projected climate changes are likely to alter the pattern of extreme weather, including tropical cyclones; hence, a greater number of severe storms are expected. The vast coastal borders, the many islands and atolls, long coastlines, and high population concentration in the coastal regions of South Pacific nations have made tsunamis, flooding, and abnormal tides the most disastrous problems in the area (Fletcher et al., 2013).

The phrase "climate change" refers to all instances of climate change, regardless of the reasons behind them, natural fluctuations, or human influence, such as CO_2 emissions, throughout the history of the planet (Alley et al., 2007). A UNICEF report published in 2010 foretells the negative consequences of climate change in the Asia Pacific region, such as increased mortality and injury associated with extreme weather; intense conflict over water; outbreaks and redistribution of diseases caused by water, food, and vectors; migration and decline in people's resources; and food insecurity and malnutrition among children (Urbano et al., 2010).

Besides, in the case of the Pacific Island Countries (PICs), many "non-climate" factors and developments weaken local populations' ability to withstand climate pressures. Such factors include, but are not limited to, overpopulation, conflicts over land rights and family ties, western influence on traditional governance, loss of traditional knowledge affecting agriculture, resource depletion, and increased dependency on imported products (Pelling & Uitto, 2001; Warrick, 2010). Moreover, the situation of small population size and economic instability was further aggravated by disasters, necessitating humanitarian aid and development support to address their consequences and the impacts of climate change (Fletcher et al., 2012; Maclellan, 2011). The Australian government, local, regional, and international nongovernmental organizations (NGOs), as well as multilateral organizations comprising United Nations agencies, among others, are the principal development partners who, in addition to providing timely humanitarian assistance, also work in the PICs. Some of these organizations also focus on transitioning from disaster response to post-disaster development programs designed to boost overall in-country capacity. In PICs, disaster preparedness, risk reduction, and adaptation to climate change are among the areas of development assistance (Fletcher et al., 2013).

A disability is a condition in which the body or mind is impaired, making it difficult to perform specific duties and interact with the external environment. According to WHO, disability has three dimensions: First, impairments that affect the structure, function, or mental capacity of the body, such as limb loss, vision impairment, or memory decline; second, activity limitations, such as trouble seeing, hearing, walking, or solving problems; and third, participation restrictions in regular engagement activities, such as work, social interaction, recreation, and access to healthcare and preventive services (Saketkoo et al., 2022). People with disabilities (PwD) have significant challenges during disasters, but preparedness measures are essential to reducing the adverse effects on their health and society. These tactics include: addressing the vulnerabilities of PwD residing in disaster-prone nations; utilizing PwD's perceptions and experiences of past disasters; removing obstacles to disaster preparedness; and preparing for disasters by using services provided to PwD during disasters, such as volunteer work, communication accessibility, and government/NGO rescue personnel. Despite attempts to prepare for disasters, PwD are frequently ignored, marginalized, discriminated against, and left behind (Elisala et al., 2020).

People with disabilities are commonly a minority group that is marginalized in most countries, and their day-to-day lives, which are already difficult due to their conditions, cannot be further worsened by the world outside. Among the significant health risks to PwD caused by climate change are water-related diseases, heat-related diseases, mental health problems, injuries, and even death. PwD, especially those in poorer countries, are the PwD most exposed to the adverse health effects of the above-mentioned natural calamities. PwD are suffering more than others as climate change is leading to a rise in natural and artificial disasters, where the death and illness rates are higher among the disabled. Moreover, disabled people are at a greater risk of having their already frail health affected, especially with the continuous occurrence and intensification of such disasters. Negative life experiences brought on by extreme weather events linked to climate change can result in harmful stress, sorrow, sadness, and mental disease in addition to impairment of physical health. Additionally, compared to the general population, PwD have more disruptions in their medical treatment due to climate change weather events, and they are more likely to experience negative economic and social risk factors, including unemployment and poverty. PwD are particularly vulnerable to extreme weather disasters, yet their perspectives and real-world experiences are frequently left out of the creation of climate change disaster management plans (Uddin et al., 2024).

In this chapter, the perspectives of disabled individuals residing in the Pacific Islands were highlighted. This chapter contributes significantly to the connection of empirical analysis with human experience by using the scientific evidence on climate disasters and the personal narratives of disabled survivors. The chapter's integrated approach has demonstrated its relevance to the more extensive discussions of inclusive resilience, sustainable development, and human rights in disaster-affected areas through the combination of scientific evidence on climate disasters and the personal stories of the disabled. The chapter reaches the point that if the voices of everyone, particularly the most marginalized, are listened to and included in policies and practices for disaster risk reduction and climate adaptation, only then will the Pacific be truly resilient.

4.2 Disability and Disasters in the Pacific Region

A disaster is defined as "a serious disruption of society's functioning that poses a significant, widespread threat to human life, health, property, or the environment, whether arising from accident, nature, or human activity, whether developing suddenly or as a result of long-term processes, but excluding armed conflict." Disasters can be categorized as "natural" or "man-made" (war, conflict, etc.) based on their etiology. The term "natural disaster" may be described as "a situation or event caused by nature that overwhelms local capacity, necessitating a request to a national or international level for external assistance; an unforeseen and often sudden event that causes great damage, destruction, and human suffering." Natural disasters can be classified as geophysical (earthquake, volcano, dry mass movement), meteorological (storms), hydrological (flood, wet mass movement), climatological (extreme temperature, drought, wildfire/bushfire), and biological (epidemic, insect infestation, animal stampede) (Khan et al., 2015; Vos et al., 2010; WHO, 2008). The factors causing the compounded risk to the disabled person are depicted in Fig. 4.1.

The WHO has systematically developed a standardized conceptualization and method for measuring impairment. WHO introduced a measure in the first GBD research that enabled comparisons of the effects of illnesses. This gave a measure of parity between mental and physical problems and, for instance, made it possible to compare depression with diabetes. The GBD research was based on a disability model that emphasized health declines. Since then, this concept has been developed in the ICF as either an inherent characteristic of the person or the result of the interplay between the person's health status and external circumstances. The World Health Organization's World Report on Disability and World Report on Ageing and Health both continue to reflect this conceptualization (Assembly, 2015; WHO, 2007, 2015, 2016).

The face validity of WHO's health measurement strategy was based on three consensus points regarding health: that health is an intrinsic characteristic of the individual; that health is a function of states or conditions of the human body or mind, constituted by the person's intrinsic capacity to execute specific tasks and actions in a range of domains that capture the full breadth of human functioning; and that health is a determinant of, but does not coincide with, well-being (Salomon et al., 2003). When considered together, health, for measurement purposes, is seen as an aggregate of intrinsic areas of functioning that characterize an individual's state of health. The population as a whole can be used to aggregate this further.

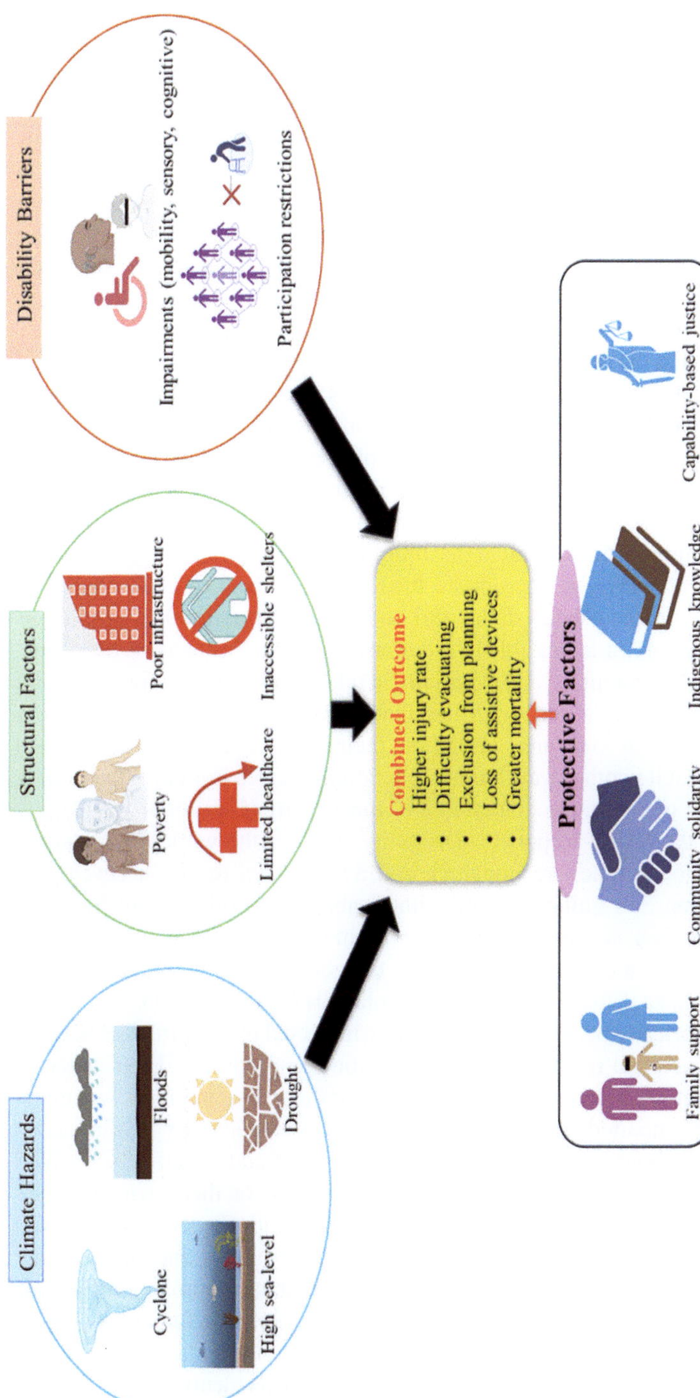

Fig. 4.1 Interaction of climate hazards, structural inequities, and disability barriers creating heightened risks and adverse outcomes for PwD. **This figure** illustrates how climate hazards, structural factors, and disability-specific barriers intersect to intensify disaster-related risks for persons with disabilities. The left panel highlights climate threats, and the central panel shows structural challenges, including poverty, inadequate infrastructure, limited healthcare access, and unsafe housing conditions. The right panel depicts disability-related barriers. Together, these overlapping factors contribute to compounded outcomes, including higher injury rates, difficulty evacuating, exclusion from disaster planning, loss of assistive devices, and increased mortality

It follows that human functioning is on a continuum, as are declines in functioning or impairment. Even though people may have quite distinct functional profiles across several domains that are indicative of their underlying medical issues, these profiles can be merged so that the degree of functioning (or impairment) can be compared across them. A shared conceptual framework enables comparisons along a single scale whenever health declines are assessed, whether through data from population health surveys or in the estimations supporting global health reporting. A set of principles form the foundation of all of WHO's current work: The first is that disability is a universal human experience rather than a characteristic of a demographic minority; however, it makes sense for activists to maintain that theirs is a minority identity given the social stigma attached to disability. Second, impairment is etiologically neutral, meaning that the decline in functioning is neither associated with nor based solely on the presence of a particular medical illness. Having trouble leaving one's house has a similar effect on a person's life, regardless of whether it is caused by a pathological dread of open areas like agoraphobia or limited movement due to a spinal cord injury. Parity between disabilities resulting from physical and mental health disorders is guaranteed by etiological neutrality. Finally, there is a spectrum of disabilities, ranging from complete disability to no disability (full functioning). A threshold determined to be appropriate for the objective, such as campaigning for policy change, can be used to divide this continuum. Disability is a continuous phenomenon; therefore, it is possible to monitor changes in this amount over time, across people and communities, and especially in relation to clinical or public health treatments. Since there is a very high likelihood that a person will experience a decline in functioning in some area throughout their life, disability is both universal and ongoing. To put it another way, human functioning ranges from full functioning to varying degrees of restriction to total loss of functioning (Cieza et al., 2018).

According to the World Report on Disability, 15% of the world's population, roughly 480 million people, are impaired, and 8% of them reside in developing nations. Every year, 350 million people with disabilities worldwide are impacted by natural catastrophes, according to estimates from the World Health Organisation (WHO). These figures are expected to rise due to ageing, lifestyle-related illnesses, poverty, and accidents (Elisala et al., 2020; Tavola, 2012; WHO, 2011).

PwD have historically been disproportionately impacted by catastrophes because of a variety of functional restrictions, social isolation, and poverty. Their demands vary depending on the nature and degree of their impairment and their functional limits. A visually impaired individual could react, for example, by hearing radio or television alerts about an approaching calamity. A deaf and cognitively disabled person could find it more challenging to make this decision on their own since they would have to rely entirely on their caretaker to help with the evacuation. PwD are more susceptible to disasters due to structural determinants, including social, political, economic, cultural, and environmental factors that may affect how people respond to disasters. The United Nations International Strategy for Disaster Risk Reduction (UNISDR) promotes risk reduction through the use of the Sendai Framework for Disaster Risk Reduction and other international agreements to lower disaster risk and vulnerability among PwD (Alexander et al., 2012; Lord et al., 2016; Smith et al., 2012; UNISDR, 2015).

Thirty percent of the world's surface is covered by the Pacific Ocean, where the PICs are widely scattered. These island nations, which are divided into "continental islands," "oceanic islands," and "raised limestone islands," differ not just geographically but also topographically (Costella & Ivaschenko, 2015). The Pacific region's PwD are among the first to suffer during catastrophes because of environmental exposure, social isolation, and the lack of disaster preparedness, and as natural disasters, which affect the whole area. The case of cyclones, tsunamis, and floods, and climate change, among others, are the main hurdles for PwD who are always the last ones to get access to early warning systems, shelters, and inclusive evacuation plans. The poor condition of the infrastructure, the communication gap, and no disability-specific policies are all contributing to their vulnerability. Studies done in the Pacific region reveal that disaster response managers hardly ever take into account the wide-ranging needs of PwD; thus, they are literally left out of the planning and recovery process. The proposals for disability-inclusive policies, the provision of accessible communication, and the consultation of PwD on all related matters are essential for justice in disaster resilience. The development of community alliances and the at-scale practice of inclusivity in the approach to disaster risk reduction will be the means through which the Pacific societies will be transformed into safer and more resilient ones, being the defenders of all, including the disabled people (Good & Phibbs, 2017; Leong et al., 2020).

The study conducted by Elisala et al. (2020) sought to understand the outlooks, readiness levels, and disaster-related incidents of disabled persons, highlighting the primary factors that make or break their disaster preparedness. The most significant factors, among other things, were the development of skills, participation in policy and community decision-making, and the availability of communication channels. Development of skills, particularly through evacuation drills, was considered vital in creating awareness of risks among persons with disabilities and the rest of the community, which is in line with the Sendai Framework that promotes the empowerment of marginalized groups through DRR activities (UNISDR, 2015; Handicap International, 2015; Elisala et al., 2020). The research, on the other hand, brought to light the issue that emergency personnel were not quite aware of the major types of disabilities among the afflicted, thus making the evacuation process quite perilous. PwD were involved in the governmental and communal decision-making processes, and their inclusion was a further argument for preparedness and support, thus correlating with the studies' viewpoint that an excellent policy involvement of PwDs empowers them and is in line with the recommendations of the International Handicap (Handicap International, 2015; Lord et al., 2016; WHO, 2013). The problem of communication was, however, still a great concern; using caregivers, local languages, sign languages, and making announcements even closer were some of the methods that greatly helped to improve understanding, which goes hand in hand with previous findings on risk.

The research pinpointed several significant barriers, such as joblessness, absence of disaster preparedness measures, and over-dependence on outside support that rendered the PwDs unready to tackle a calamity (CDC, 2017; FEMA, 2014; WHO, 2013). Over 90% of the respondents were not employed, and this clearly pointed to the broader socioeconomic issue as well as the nonavailability of the most basic resources (Alexander et al., 2012; Botzen et al., 2009; ILO, 2014, 2015; WHO, 2013). Creating more job opportunities and implementing inclusive policies were considered the main points to focus on, as they would lead to greater resilience among the affected population. The factors that prompted people to prepare for a disaster were survival instincts, Te Valo (local alert systems), and faith, which originated in Pacific cultural traditions and were still very much alive. Information sources that people trusted, government agencies, and community alerts were the most important ones, and contributed greatly to preparedness, although there was a reduction in trust due to delayed communication. Moreover, both scientific and traditional knowledge were found to be essential in determining preparedness behaviors, while personal experiences were the central factor influencing risk perception and coping strategies. The results of the study stress that the different experiences and the knowledge of the PwD should be taken into account in disaster planning; thus, inclusive, participatory, and tailored approaches have to be used to make resilience in their communities stronger (Clark et al., 1993; Dekens, 2007a, 2007b; Dorasamy et al., 2013; Elisala et al., 2020; Gaillard & Texier, 2010; Helweg-Larsen, 1999; Hiwasaki et al., 2014; Karanci & Aksit, 1999; McAdoo et al., 2006, 2009; Steelman & McCaffrey, 2013).

4.3 Firsthand Accounts of Disabled Individuals During Disasters

A study was carried out by Anesi in the year 2022 on the tragic case of Hans Dalton, a man from Samoa with a mental disability who died a sad and unfortunate death during Cyclone Evan (2012). This is a testimony of the ableism, colonialism, and institutional neglect that characterized the crisis response to disabled people in the Pacific region. Cyclone Evan (2012) was the natural calamity that changed this patient's fate into an accidental medical disaster. The poor fellow, confined in a hospital, then after a police station, and finally in a prison, where the last stop was already under control and cut off from colonial areas, were the last places in his sorrowful journey when the cyclone caused the loss of his medication. The case of his death has been officially labeled as drowning, but at the same time, it is suggestive of possible physical trauma, and it reveals that mental disabilities are still very often criminalized in post-colonial Samoa rather than treated with humanity as would be expected. The response of institutional actors to Dalton's suffering cast a clear light on the persistence of such colonial hierarchies in the healthcare and justice sectors, where Indigenous disabled people are considered ill, monitored, and punished. This tale shows how people with mental disabilities are always the first among victims to nature's vile and violent hands, not only by the natural threats to the environment but also by the cruelty and neglect of the system. The acknowledgment of such tales is necessary to alter disaster preparedness strategies to include culture-based, anti-ableist, and community-driven care for disabled men and women in the Pacific Ocean (Anesi, 2022).

The main stories of the iTaukei, indigenous Fijians of Votua and Navala, both located in the Ba River catchment of Fiji, uncovered through a study of Cyclone Winston (2016), clearly show the massive impact of disability, gender, and social class on the entire process of disasters and coping. The respondents were very straightforward in saying that the vulnerability tag was more closely linked with different physical limitations, disability, age, and loss of house than with any of the genders. They gave in when someone talked about a woman from Votua who, for the sake of the mobility-impaired brother-in-law, had to choose to postpone the living care of him, as evacuating to the centers was a tough deal, plus the area was that crowded. She further indirectly said that physical disability increased the vulnerability to disaster effects. Similarly, older adults and people with disabilities consistently ranked as the most affected, unable to reach a safe or shelter place beforehand due to the flood covering the area (FAO, 2016; Griffen, 2006). These testimonies of the people revealed the preparedness and resilience undertaken by the Fijian communities, which are in-depth intertwined with social cohesion, caregiving responsibilities, and communal reciprocity, where individuals with disabilities are forced to rely on the support systems offered by family and neighbors. The narratives also affirmed that firsthand lived experiences, especially with disabled people, are vital for developing inclusive, culturally grounded DDR frameworks that foreground embodied, context-specific resilience rather than homogenized assumptions about vulnerability (Bennett et al., 2020).

According to the findings of the Tropical Cyclone Pam (Vanuatu, 2015) study, the disability community suffered the most tremendous impact from the disaster, and the main reasons were physical accessibility problems, exclusion from the preparedness process, and the absence of support systems during and after the disaster. Consequently, many disabled people lived in disaster-affected areas, or their mobility and communication capacities were such that they could not get to the shelters for evacuation, or they were only getting through food and health services. The study reported that PwD were 2.45 times more likely to be injured than non-disabled persons, and the humanitarian response teams were very often not trained or equipped with the means to meet disability-specific needs at their location. The cyclone not only cut off the supply of vital services for disabled people but also took away the devices meant for mobility, hence leading to social isolation. However, the research found that when PwD were part of community disaster committees and assigned leadership roles, the community's preparedness and resilience were surprisingly boosted. The report pinpoints that disability inclusion should be integrated into the three stages of humanitarian action (i.e., preparedness, response, and recovery), and it pointed out that the full involvement of persons with disabilities is equivalent to the empowerment of communities that are both resilient and fair (Baker et al., 2017).

One of the most significant cases among the disabled in the Solomon Islands during the floods was that of Ellena, who underwent leg amputation. Ellena narrated how the floods restricted her movement drastically because her crutch was either lost or broken. Mobility became an absolute issue for her, and she was completely helpless and thus reliant on others to move around. She further gave an account of how, at times, people with disabilities were the last ones to be evacuated or, in some instances, even not evacuated at all, which, to her, was proof of the planners' ignorance and neglect in this area of disaster response planning. Her account indicates that lack of proper infrastructure, escape routes that are not accessible, and negative social attitudes towards disability are all factors that increase the risk during crises. But Ellena does not give up; she is a regular participant in community outreach programs and a strong advocate for disability-inclusive disaster risk reduction (DiDRR), which, in fact, is a demonstration of the fact that PWDs can be resilient if they are allowed to be part of local preparedness initiatives through their participation and Empowerment. Her account highlights the need for inclusive community-based disaster planning, which prioritizes accessibility and the participation of PWDs at every stage of response and recovery (CBM Australia, 2022; Fuhrer, 2014).

4.3.1 Survival and Adaptation Strategies

In Pacific Island societies, characterized by kinship and communal solidarity as the basic elements of social life, family and community ties are the major factors that help PwD survive disasters. Without structured institutional support, families tend to be the first to respond during such situations and help disabled members with evacuations, shelters, and post-disaster recovery. Research conducted in Fiji, Samoa, and Vanuatu has demonstrated that PwD depend heavily not only on their close relatives but also on their neighbors and traditional community structures for physical assistance, mobility support, and emotional care. The way in which the Pacific peoples show love through veiwekani (kinship) and alofa (care/love) embodies the same resilience, which is community-based, and thereby gives the most vulnerable, disabled people among them, rarely complete isolation. Nevertheless, these ties, on the one hand, create a solid social protection, while, on the other hand, they might conceal the lack of care within the system when the responsibilities of caregiving are placed entirely on families with no help from institutions or the government (Aipira et al., 2017; FAO, 2016).

Adaptation has become an urgent need for survival for several South Pacific cultures. Over the past several years, there has been an increasing need to adapt to the challenges posed by climate change. Initially, around half of Fiji's estimated 902,964 people receive significant benefits from the country's marine and coastal environments, including physical, economic, social, ecological, and cultural benefits. However, the way of life of coastal people is threatened by the effects of climate change on coastal ecosystems, and for the population of Vunidogoloa in the province of Cakaudrove in Vanua Levu, relocation has been a reality for over 30 years. Relocation may be the final option, but it is also one of the best adaptation techniques for many coastal Fijian villages now facing struggles similar to those in Vunidogoloa.

This implies that many of these disadvantaged groups could face challenges similar to those faced by Vunidogoloa inhabitants (Charan et al., 2017).

A capabilities-based approach to climate justice can be seen as a radical new way of looking at climate change and its impact on people. It is all about the conditions—environmental, sociocultural, and developmental—that must be provided for individuals and communities not only to survive but also to function and grow in a world that is slowly changing climate-wise. As the authors Wolff and de-Shalit note, disadvantages are often interlinked, which means that a person losing one capability is likely to lose others as well; climate change thus operates as a "corrosive disadvantage" that not only enhances inequalities that already exist but also puts the most vulnerable groups at risk of heightened impacts like droughts, floods, food shortages, health hazards, and displacement. The task of the state is, accordingly, to dissolve these disadvantage clusters and set up "fertile functionings," where reinforcing one capability will also contribute to developing the others. This, in turn, entails being able to recognize, illustrate, and prioritize vulnerabilities connected with climate change by utilizing both scientific and local knowledge. Vulnerability mapping has to be community-based and not top-down; thus, it is necessary to have the active involvement of the local population in the process so that the resulting adaptation policies could truly reflect the realities of life and also acknowledge the principles of recognition and democratic participation. It is through this comprehensive and location-specific method that policymakers will be able to determine what human capabilities are essential to protect, restore, and assess the effectiveness of climate change adaptation strategies, and still deliver justice in an already altered climate world (Schlosberg, 2012; Vallentyne, 2007).

The impact of climate change on gender is significant. The planning of DRR and CCA projects must consider these issues and the underlying factors of women's poverty, sustainable development, and climate disasters. The international treaties are now recognizing these links. The Paris Agreement on Climate Change, featured at the UN Framework Convention on Climate Change (UNFCCC) Conference of the Parties (COP21), demands a "country-driven, gender-responsive, participatory, and fully transparent approach to fostering climate resilience and reducing vulnerability." In the same way, the Sendai Global Framework for Disaster Risk Reduction declared that "the role of women and their participation are fundamental to the proper handling of disaster risk and to the creation of gender-sensitive disaster risk reduction policies, plans and programs." Moreover, it insisted on "the introduction of a gender, age, disability, and cultural standpoint into every policy and practice." The aims of the Sendai Framework and the results of COP21 have made the case for considering the gender aspect of climate change as a point of intervention stronger than ever (Aipira et al., 2017).

PwD presence during catastrophes is largely dependent upon one main aspect: access to information. Barriers in communication, like a lack of sign language interpreters, inaccessible alerts, and notifications for non-participants via articles, can all lead to the same consequence of delaying evacuations and making the situation worse. The use of inclusive communication systems, e.g., audio-visual alarms, tactile notifications, and mobile warning apps with accessibility features, is not only a movement but a major step forward for disaster response in the Pacific. By the same token, local-language radio-based warnings in Vanuatu connect the impaired visually as well as aurally to the information. The changes in technology being advanced, together with the training of caregivers and local responders, ensure that the essential information is accessible to everyone regardless of their disabilities. It is a common misconception to think that inclusive communication is a luxury service; rather, it should be acknowledged as a major stake in the battle for equal disaster preparedness (Delforge et al., 2025; Hamidzada & Cruz, 2017; WHO, 2022).

4.4 Building Inclusive Resilience: Lessons from Islands

Disability Forum (PDF) and Handicap International have been actively involved in disability mapping, community awareness-raising, and skill development for persons with disabilities, thereby enhancing their preparedness and empowerment (CBM Australia, 2022; Handicap International, 2015). The governments of the region, with the help of UNDP and WHO, have started to integrate disability indicators in their national disaster strategies. There are still, however, lots of gaps, particularly in poor resource areas where the implementation of the strategies is not uniform (Fletcher et al., 2013). The Sendai Framework, CRPD, and the Paris Agreement are illustrations of international agreements that furnish a normative structure that makes a point of accessibility, participation, and equity in DRR being the very heart of the matter. Therefore, a collaborative, multi-level governance approach is a must for the Pacific to become a region of resilient, inclusive communities.

The frameworks that recognize the long-term effects of disasters on people with disabilities and guarantee that rehabilitation, reconstruction, and livelihood programs are accessible to every new and older adult have to be used for the Sustainable and inclusive recovery. The capability approach, which focuses on the reduction of human capabilities due to climate change and natural disasters, is an ideal way to set up recovery programs that aim to rebuild infrastructure and social dignity simultaneously (Schlosberg, 2012). The Sendai Framework (2015–2030) has proposed a new approach to development that integrates a disability perspective and seeks to reduce systemic inequalities, which are among the primary obstacles to the realization of development goals. The communities in the Pacific who are working towards inclusive recovery are also adhering to the tram to be close to the earth, practice kinship, and share, with their indigenous roots traced back to the principles of community care (Aipira et al., 2017). The global principles of DRR would not just help us in the battle against disasters but also help us to get a better understanding of the local contexts and experiences, and therefore the outcome of the rebuilding process would not be the same with the past situation in terms of vulnerability but would rather be the resilience of the affected persons with disabilities for the long term.

When people with disabilities are given the role of leaders in disaster preparedness and climate adaptation, it will not only be an advantage for the individual group but also a stepping stone to the goal of long-term inclusive resilience. The engagement should not only be through consulting but rather involve prominent positions like village councils, disaster risk-reduction committees, and national disaster management agencies (Lord et al., 2016). It has been observed through research that inclusion of persons with disabilities allows the entire community to come up with more practical and safe evacuation plans, easier communication methods, and culturally appropriate preparedness practices (Elisala et al., 2020). The outlay on strengthening human resources, leadership, and accessibility of meetings will make participation equal and thus a requirement for the empowerment of persons with disabilities. Moreover, the partnerships between disability advocacy groups, customary leaders, and the government will not only raise the profile of persons with disabilities but also improve the whole community's readiness to respond. Ultimately, the disaster governance of persons with disabilities transforms their image from that of a burden on the community to that of a resilient and strong community that has the potential to support and empower the disability group.

4.5 Conclusion

Instances elucidating the theme of the chapter unequivocally manifested that the Pacific Island Countries, which are undeniably disturbed by natural calamities most frequently, are also the places where the problems faced by disabled persons are experienced to the fullest extent. People with disabilities are not just suffering from their physical limitations but also from the combination of social injustice, unpreparedness of the system, stigma, plus the persistence of colonial structures. The heartbreaking personal testimonies—ranging from the death of Hans Dalton in Samoa to the plight of disabled survivors in Fiji, Vanuatu, and the Solomon Islands—have undoubtedly pointed to the necessity of persons with disabilities being involved throughout the different stages of disaster management. However, these stories also hint at the power, endurance, and talent that PWDs offer to their society through the use of artistic forms of communication, participation in DRR committees, advocacy for open communication, and a strong reliance on family support networks. The survival and adaptation of the Pacific are processes of the whole community shaped by the Indigenous peoples' values of reciprocity, care, and communal responsibility. If the right policies that improve accessibility, democratic participation, capability protection, and culturally appropriate adaptation are in place, People with Disabilities can be very sociable and make the community strong against future disasters. That is to say, a really strong Pacific region would first need the complete dismantling of all ableist and exclusionary practices, secondly, the merging of scientific and traditional knowledge, and thirdly, the placement of disabled voices right in the middle—not on the outskirts—in disaster preparedness and climate adaptation. Only with inclusive, fair, and community-led strategies can the region's climate resilience prevail, as the danger of global warming keeps on intensifying.

References

Aipira, C., Kidd, A., & Morioka, K. (2017). Climate change adaptation in Pacific countries: Fostering resilience through gender equality. In *Climate change adaptation in Pacific countries: Fostering resilience and improving the quality of life* (pp. 225–239). Springer International Publishing. https://doi.org/10.1007/978-3-319-50094-2_13

Alexander, D., Gaillard, J. C., & Wisner, B. (2012). Disability and disaster. In *Handbook of hazards and disaster risk reduction* (pp. 413–423). Routledge. https://www.researchgate.net/publication/285004501_Disability_and_disaster

Alley, R. B., Berntsen, T., Bindoff, N. L., Chidthaisong, A., Friedlingstein, P., Gregory, J. M., Hegerl, G. C., Heimann, M., Hewitson, B., Hoskins, B. J., Joos, F., ... & Miller, H. L. (2007). Summary for policymakers. *Climate change 2007: The physical science basis. contribution of working group I to the Fourth assessment report of the intergovernmental panel on climate change* [Solomon, S., D. Qin, M. Manning, Z. Chen, M. Marquis, KB Averyt, M. Tignor and HL Miller (eds.)].

Anesi, J. (2022). Enduring the storm: Dealing with mental disabilities in Oceania. *Disability Studies Quarterly, 41*(4). https://doi.org/10.18061/dsq.v41i4.8457

Assembly, U. G. (2015). *Transforming our world: The 2030 Agenda for sustainable development.* https://sdgs.un.org/2030agenda

Baker, S., Reeve, M., Marella, M., Roubin, D., Caleb, N., & Brown, T. (2017). Experiences of people with disabilities during and after Tropical Cyclone Pam and recommendations for humanitarian leaders. In *Asia Pacific Humanitarian Leadership Conference Proceedings* (pp. 70–78). https://ojs.deakin.edu.au/index.php/aphl/article/view/825

Bennett, K., Neef, A., & Varea, R. (2020). Embodying resilience: Narrating gendered experiences of disasters in fiji. In *Climate-induced disasters in the Asia-Pacific region: Response, recovery, adaptation* (Vol. 22, pp. 87–112). Emerald Publishing Limited. https://doi.org/10.1108/S2040-726220200000022004

Botzen, W. J., Aerts, J. C., & van den Bergh, J. C. (2009). Dependence of flood risk perceptions on socioeconomic and objective risk factors. *Water Resources Research, 45*(10). https://doi.org/10.1029/2009WR007743

Bündnis EH (2012). *World Risk Report 2012.* Berlin. https://weltrisikobericht.de/wp-content/uploads/2016/08/WorldRiskReport_2012.pdf

CBM Australia. (2022). *Disability-inclusive disaster risk reduction in the Solomon Islands: Ellena's story.* CBM Australia. https://www.cbm.org.au/stories/disability-inclusive-disaster-risk-reduction-solomon-islands-ellena-story

CDC (2017). *Emergency preparedness: Including people with disabilities: Centre for diseases control and prevention.* http://www.cdc.gov/ncbdd/disabilityandhealth/emergencypreparedness.html

Charan, D., Kaur, M., & Singh, P. (2017). Customary land and climate change induced relocation—A case study of Vunidogoloa Village, Vanua Levu, Fiji. In *Climate change adaptation in Pacific Countries: Fostering resilience and improving the quality of life* (pp. 19–33). Springer International Publishing. https://doi.org/10.1007/978-3-319-50094-2_2

Cieza, A., Sabariego, C., Bickenbach, J., & Chatterji, S. (2018). Rethinking disability. *BMC Medicine, 16*(1), 14. https://doi.org/10.1186/s12916-017-1002-6

Clark, L. V., Venezlano, L., & Atwood, D. (1993). Situational and dispositional determinants of cognitive and affective reactions to the New Madrid earthquake prediction. *International Journal of Mass Emergencies & Disasters, 11*(3), 323–333. https://doi.org/10.1177/028072709301100305

Costella, C., & Ivaschenko, O. (2015). Integrating disaster response and climate resilience in social protection programs in the Pacific Island countries. *Social Protection and Labour World Bank, 1*(1507), 1–68. https://openknowledge.worldbank.org/server/api/core/bitstreams/cbeedc81-e1c1-57b3-b75d-4ddb6bb810ba/content

Coumou, D., & Rahmstorf, S. (2012). A decade of weather extremes. *Nature Climate Change, 2*(7), 491–496. https://doi.org/10.1038/nclimate1452

Dekens, J. (2007a). *The river and the snake don't run straight, local knowledge on flood preparedness in the Eastern Terai of Nepal*. Kathmandu: ICIMOD. https://doi.org/10.53055/ICIMOD.481

Dekens, J. (2007b). *Herders of Chitral, the lost messengers?: Local knowledge on disaster preparedness in Chitral District, Pakistan*. International Centre for Integrated Mountain Development. https://doi.org/10.53055/ICIMOD.469

Delforge, D., Wathelet, V., Below, R., Sofia, C. L., Tonnelier, M., van Loenhout, J. A., & Speybroeck, N. (2025). EM-DAT: The emergency events database. *International Journal of Disaster Risk Reduction,* 105509. https://doi.org/10.1016/j.ijdrr.2025.105509

Dorasamy, M., Raman, M., Marimuthu, M., & Kaliannan, M. (2013). Disaster preparedness: An investigation on motivation and barriers. *Journal of Emergency Management (Weston, Mass.), 11*(6), 433–446. https://doi.org/10.5055/jem.2013.0156

Elisala, N., Turagabeci, A., Mohammadnezhad, M., & Mangum, T. (2020). Exploring persons with disabilities preparedness, perceptions and experiences of disasters in Tuvalu. *PLoS ONE, 15*(10), Article e0241180. https://doi.org/10.1371/journal.pone.0241180

FAO. (2016). *Fiji: Tropical Cyclone Winston: Situation report—16 March 2016*. Rome: FAO. http://www.fao.org/resilience/resources/resources-detail/en/c/396283/

FEMA. (2014). *Preparedness in America: Research insight to increase individual, organization and community action: Federal Emergency Management Agency*. https://www.fema.gov/media-library-data

Fletcher, S. M., Thiessen, J., Gero, A., Rumsey, M., Kuruppu, N., & Willetts, J. (2013). Traditional coping strategies and disaster response: Examples from the South Pacific region. *Journal of Environmental and Public Health, 2013*(1), Article 264503. https://doi.org/10.1155/2013/264503

Fletcher, S., Gero, A., & Willetts, M. (2012). *Review of Australia's overseas disaster and emergency response sector*. Prepared for NCCARF, WHO Collaborating Centre and the Institute for Sustainable Futures, University of Technology, Sydney, Australia.

Food and Agriculture Organization (FAO). (2016). *Cyclone Winston: Recovery and rehabilitation needs assessment*. Rome: FAO. https://www.fao.org/fileadmin/user_upload/emergencies/docs/FAOSitRep_FijiTropicalCycloneWinston_140416.pdf

Fuhrer, M. (2014). Disability inclusive disaster risk reduction. *Planet@ Risk, 2*(3).

Gaillard, J. C., & Texier, P. (2010). Religions, natural hazards, and disasters: An introduction. *Religion, 40*(2), 81–84. https://doi.org/10.1016/j.religion.2009.12.001

Good, G. A., & Phibbs, S. (2017). Disasters and disabled people: Have any lessons been learned? *Journal of Visual Impairment & Blindness, 111*(1), 85–87. https://doi.org/10.1177/0145482X1711100109

Griffen, V. (2006). Local and global women's rights in the Pacific. *Development, 49*(1), 108–112. https://doi.org/10.1057/palgrave.development.1100223

Hamidzada, M., & Cruz, A. M. (2017). Understanding women's vulnerability factors to natural hazards in Afghanistan. *B, 60*(B), 343–349.

Helweg-Larsen, M. (1999). (The lack of) optimistic biases in response to the 1994 Northridge earthquake: The role of personal experience. *Basic and Applied Social Psychology, 21*(2), 119–129. https://doi.org/10.1207/S15324834BA210204

Hiwasaki, L., Luna, E., & Shaw, R. (2014). Process for integrating local and indigenous knowledge with science for hydro-meteorological disaster risk reduction and climate change adaptation in coastal and small island communities. *International Journal of Disaster Risk Reduction, 10*, 15–27. https://doi.org/10.1016/j.ijdrr.2014.07.007

ILO. (2014). Facts on Disability in the world of work. In Organization IL, editor. Geneva: ILO.

ILO. (2015). Facts on disability in the world of work. In Organization IL, editor. Geneva: ILO.

International. H. Annual Report. (2015). Lyon Cedex, France: Handicap International.

Karanci, N. A., & Aksit, B. (1999). Strengthening community participation in disaster management by strengthening governmental and non-governmental organisations and networks: A case study from Dinar and Bursa (Turkey). *The Australian Journal of Emergency Management, 13*(4), 35–39.

Khan, F., Amatya, B., Gosney, J., Rathore, F. A., & Burkle, F. M., Jr. (2015). Medical rehabilitation in natural disasters: A review. *Archives of Physical Medicine and Rehabilitation, 96*(9), 1709–1727. https://doi.org/10.1016/j.apmr.2015.02.007

Leong, G. M., Stodden, R., & Balikuddembe, J. K. (2020). *Disability and disaster: What to know-what to do.*

Lord, A., Sijapati, B., Baniya, J., Chand, O., & Ghale, T. (2016). *Disaster, disability, & difference.* https://doi.org/10.13140/RG.2.2.30295.93609

Maclellan, N. (2011). *Turning the tide: Improving access to climate financing in the Pacific Islands.*

McAdoo, B. G., Dengler, L., Prasetya, G., & Titov, V. (2006). Smong: How an oral history saved thousands on Indonesia's Simeulue Island during the December 2004 and March 2005 tsunamis. *Earthquake Spectra, 22*(3_suppl), 661–669. https://doi.org/10.1193/1.2204966

McAdoo, B. G., Moore, A., & Baumwoll, J. (2009). Indigenous knowledge and the near field population response during the 2007 Solomon Islands tsunami. *Natural Hazards, 48*(1), 73–82. https://doi.org/10.1007/s11069-008-9249-z

Pelling, M., & Uitto, J. I. (2001). Small island developing states: Natural disaster vulnerability and global change. *Global Environmental Change Part b: Environmental Hazards, 3*(2), 49–62. https://doi.org/10.3763/ehaz.2001.0306

Saketkoo, L. A., Escorpizo, R., Varga, J., Keen, K. J., Fligelstone, K., Birring, S. S., ... & Global Fellowship on Rehabilitation and Exercise in Systemic Sclerosis (G-FoRSS). (2022). World Health Organization (WHO) international classification of functioning, disability and health (ICF) core set development for interstitial lung disease. *Frontiers in pharmacology, 13*, 979788. https://doi.org/10.3389/fphar.2022.979788

Salomon, J. A., Mathers, C. D., Chatterji, S., Sadana, R., Ustun, T. B., & Murray, C. J. (2003). Quantifying individual levels of health: definitions, concepts and measurement issues. *Health Systems Performance Assessment: Debate, Methods, and Empiricism*, 301–18.

Schlosberg, D. (2012). Climate justice and capabilities: A framework for adaptation policy. *Ethics & International Affairs, 26*(4), 445–461. https://doi.org/10.1017/S0892679412000615

Smith, F., Jolley, E., & Schmidt, E. (2012). *Disability and disasters: The importance of an inclusive approach to vulnerability and social capital.* Haywards Heath: Sightsavers.

Steelman, T. A., & McCaffrey, S. (2013). Best practices in risk and crisis communication: Implications for natural hazards management. *Natural Hazards, 65*(1), 683–705. https://doi.org/10.1007/s11069-012-0386-z

Tavola, H. (2012). *Addressing inequalities: disability in Pacific island countries.* Paper presented online at the 'Addressing inequalities' Global Thematic Consultation.

Uddin, T., Tasnim, A., Islam, M. R., Islam, M. T., Salek, A. K. M., Khan, M. M., Gosney, J., & Haque, M. A. (2024). Health impacts of climate-change related natural disasters on persons with disabilities in developing countries: A literature review. *The Journal of Climate Change and Health, 19*, 100332. https://doi.org/10.1016/j.joclim.2024.100332

United Nations Office for Disaster Risk Reduction (UNISDR). (2015). *Sendai Framework for disaster risk reduction 2015–2030.* https://www.undrr.org/publication/sendai-framework-disaster-risk-reduction-2015-2030

Urbano, M., Maclellan, N., Ruff, T., & Blashki, G. (2010). *Climate change and children in the Pacific Islands.* Nossal Institute for Global Health, University of Melbourne. https://library.sprep.org/sites/default/files/641.pdf

Vallentyne, P. (2007). Jonathan Wolff and Avner De-Shalit. *Disadvantage.* https://doi.org/10.1017/S0953820809990306

Vos, F., Rodriguez, J., Below, R., & Guha-Sapir, D. (2010). Annual disaster statistical review 2009: The numbers and trends. *Annual disaster statistical review 2009: The numbers and trends*, 46–46. http://www.cred.be/sites/default/files/ADSR_2009.pdf

Warrick, O. (2010). *Climate change and social change: Vulnerability and adaptation in rural Vanuatu.* The University of Waikato, Hamilton, New Zealand. https://pacific-data.sprep.org/system/files/102_1.pdf

World Health Organization (2016). *About the Global Burden of Disease (GBD) project.* http://www.who.int/healthinfo/global_burden_disease/about/en/

World Health Organization. (2007). *International classification of functioning, disability, and health: Children & youth version: ICF-CY.* https://www.who.int/standards/classifications/international-classification-of-functioning-disability-and-health

World Health Organization. (2008). *Glossary of humanitarian terms.* https://reliefweb.int/report/world/reliefweb-glossary-humanitarian-terms-enko

World Health Organization. (2011). World report on disability. In *World report on disability* (pp. 24–24). https://www.who.int/teams/noncommunicable-diseases/sensory-functions-disability-and-rehabilitation/world-report-on-disability

World Health Organization. (2013). *Guidance note on disability and emergency risk management for health.* https://www.who.int/publications/i/item/guidance-note-on-disability-and-emergency-risk-management-for-health

World Health Organization. (2015). *World report on ageing and health.* https://www.who.int/publications/i/item/9789241565042

World Health Organization. (2022). *Global report on health equity for persons with disabilities.* https://www.who.int/teams/noncommunicable-diseases/sensory-functions-disability-and-rehabilitation/global-report-on-health-equity-for-persons-with-disabilities

Chapter 5
Women, Disability, and Double Vulnerability

Abstract Women in the Pacific region with disabilities face discrimination that takes on different forms and is interrelated. This discrimination is mainly caused by gender inequality, cultural stigma, and precarious land rights, coupled with the socio-economic barriers that are common in the region. The chapter concentrates on a study of the combined impact of disability and gender on women who are called "double vulnerable" and charts the discussion through land inheritance and ownership problems in the Solomon Islands, the seclusion of women in rural Pacific areas, and the calamities that cause additional hardships. The authors draw on a variety of sources, including empirical research, regional statistics, and intersectional theory, to claim that women's lack of power and participation is a result of the existing norms of discrimination, low-quality education, and lack of access to good jobs, sexual and reproductive health services, and support networks. Furthermore, women with intellectual disabilities who also suffer from menstrual stigma, who do not have proper washing and sanitation facilities, and who live in areas without safe evacuation routes are the ones who suffer the most during such calamities. The underlying analysis leads to the requirement for disability-inclusive and gender-responsive policies, strong rights-based community support systems, and development approaches consistent with the CRPD, CEDAW, and the Sendai Framework. The chapter illustrates a situation in which the essential steps are to enact structural and policy reforms, upgrade data systems, and build partnerships that recognize women with disabilities as targets of the issue and as solutions to the resilience and development of the Pacific.

Keywords Women with disabilities · Pacific region · Socio-cultural determinants · Disability inclusion · Disaster risk reduction

S. M. Thomas and R. Veerabathiran, *Disability, Disaster, and Resilience in Oceania*, SpringerBriefs in Modern Perspectives on Disability Research,
https://doi.org/10.1007/978-981-95-8535-9_5

5.1 Introduction

Over 800 million elderly individuals are expected worldwide by 2025. Most of these elderly individuals will be women, and two-thirds will reside in developing nations. According to estimates from the World Health Organisation, there are 600 million disabled persons in the world, and 80% of them reside in developing nations. Up to 20% of women worldwide are thought to be disabled women and girls. People with impairments are more likely to be cared for by women. Women with disabilities, especially those from impoverished rural areas, live in extraordinary subservience, have very little influence over their lives, and experience violence and prejudice due to both their condition and their gender. Development efforts, assistance programs, poverty alleviation techniques, and human rights measures generally do not prioritize people with disabilities, and women with disabilities in particular. When they are recognized as a specific group of targets, strategies that would address underlying injustices, such as gender inequality and the realization of fundamental human rights, are largely neglected in favor of rehabilitation, impairment prevention, health care, and the provision of technical aids and equipment. Additionally, policies intended to advance women's rights and gender parity do not include women with disabilities. Disability is a significant problem for the women's rights movement, just as gender is a major one for the disability movement (Guilbert, 1999; Hans & Patri, 2003).

Similar links between gender and disability are still mostly unseen, although women's rights activists and groups have acknowledged to some degree how gender and race interact to generate specific areas of disadvantage for women. This implies that the unique experiences of women with disabilities are still largely ignored. Women with disabilities are excluded from women's organizations; many of them are unable to participate in regular events, including conferences, meetings, consultations, information sharing, and support services. Women with disabilities are not included in or reached by the policies, reports, advocacy initiatives, and development programs that seek to address women's rights (Sands, 2005).

The goal of development is to lessen social inequality and integrate marginalized groups—such as disabled women and children—into society so they may go to school, work, have children, establish a family, receive healthcare, and therapy, participate in political parties, visit theatres and places of worship, board buses, take phone calls, and use the Internet—just like any other citizen. Neither rights nor access to "social and medical" services are the only things desired. People with disabilities and their families desire and deserve their due role as citizens, just like everyone else in their community. Despite their desire to be socially and economically responsible, they too want and deserve to break free from the poverty-disability cycle that they frequently encounter in their fight for basic existence. Some strategists contend that because individuals with disabilities fall under the category of "vulnerable population" or are included in programs tailored to various genders and children, these concerns are already "mainstreamed." Disability advocates and other development professionals understand, however, that merely mainstreaming disability into current sector activities is insufficient. Experience has unequivocally shown that persons with disabilities are still marginalized in society, are mostly mute and invisible, and are therefore unable to function, communicate, and participate. One obvious instance of how handicapped women's needs have not been met is gender programming. Simply classifying individuals with impairments as "vulnerable persons" and assuming they would gain from mainstream programs is insufficient. Like women and men, children, older adults, and migrants, they are not a "homogeneous" population. People with disabilities also have distinctive needs, much as "gender and development" and "child rights" initiatives focus on the problems particular to these groups (Maya Dhungana, 2006; Sands, 2005).

The term "vulnerable" is frequently applied to women and women with disabilities. The idea of the "vulnerable female subject," which is imposed from the outside, frequently affects them. However, this externally imposed idea frequently differs significantly from the reality of how women and disabled women perceive vulnerability. Furthermore, the idea of the "vulnerable female subject" has the potential to perpetuate marginalization and restrict the options available to women and women with disabilities. Additionally, this labelling of vulnerability creates a power dynamic that benefits those not classified as susceptible.

Additionally, this may restrict the options and independence of vulnerable populations, like women and people with disabilities (Boyle, 2003; Butler et al., 2016; Hollander, 2002; Lajoie, 2018; Luna, 2009; Scully, 2014). A person must have their legal competence recognized for their autonomy to be valued on an equal footing with others. A person's legal personhood and agency are referred to as their legal capability. Women and women with disabilities are more likely to suffer denials of legal capacity and obstacles to legal capacity, such as guardianship imposition, denial of reproductive choice, and denial of decision-making in psychiatric settings. This is partly due to the term "vulnerability," which can lead to the presumption that more protection is required. Due to the severely disempowering character of autonomy elimination that accompanies constraints on legal capacity, these denials and barriers to legal capacity increase vulnerability among these populations (Arstein-Kerslake, 2019). Moreover, disabled females undergo social bias and discrimination, which are tightly interwoven with society, and this could be aggravated by negative gender stereotypes that see their disability as an indication of not being able to conform to the norms of being a wife or a mother, thereby making them more susceptible to violence (Lowe et al., 2025).

The research of Moodley and Graham (2015) very convincingly argues that disability and gender converge to create a situation of serious disadvantage, especially for Black women with disabilities. Generally, disabled women are the least educated, have no or minimal economic activity, and receive the lowest income. Besides, the intersection of race complicates the situation even more: For instance, in South Africa, a Black man without any disability might end up in a worse situation compared to a White woman with a disability with respect to education and job opportunities, which indicates that the effect of racial inequality is still very powerful. The combined structural disadvantages in the labor market continue to confine women with disabilities. They are not only the ones stripped of fundamental rights because of their gender but also the ones who are assigned family care the majority of the time, experience the greatest financial difficulties, and are further restricted due to their disability. People with disabilities already suffer from low employment rates, and when this is combined with higher travel and healthcare costs, it could lead to withdrawal from the labor market. While social grants provide the necessary support and help a significant number of women with disabilities to step out of the vicious circle of poverty, these grants rarely change people's lives; transformed and disabled ones are the least because they have the most necessary expenses to cover with the help of grants (Moodley & Graham, 2015).

Consequently, the results point to the necessity of a cross-sectional approach to the inequality issue. Race is still the most critical factor determining education, employment, and income, and when it is combined with gender and disability, it forms a triple burden that affects Black women with disabilities the most. The current policies tackling disability, gender, and race do not reach this group effectively, and thus there is a need for interventions that are specifically designed to meet the confronting inequalities in this area (Moodley & Graham, 2015).

Taking into account all these facts, this chapter scrutinizes the subtle dynamics of gender and disability as a trinitarian factor that banishes women in Oceania. It focuses on socio-cultural, structural, and economic features of women's lives and particularly highlights rural isolation, caregiving burdens, limited access to services, and increased vulnerability to violence. The chapter also discusses the changing global and regional policy contexts, with the CRPD, CEDAW, and the Sendai Framework as the primary focus, to assess the extent to which they fulfill or neglect the needs of women with disabilities. The chapter is a thorough examination bolstered by empirical evidence, case studies, and intersectional theory, and it unearths the key difficulties that women with disabilities face. It also highlights the weaknesses of existing development and disaster risk reduction approaches, leading to calls for inclusive, gender-responsive, and disability-sensitive interventions. By suggesting a radical change that recognizes women with disabilities as rights holders and resilience co-creators in the Pacific Region, the chapter finally asserts that such an approach is necessary.

5.2 Double Vulnerability: The Socio-Cultural Determinants of Disability

Disability is a complex and diversified phenomenon. It changes due to a variety of factors such as social setting, environment, and personal characteristics. The person's living condition can have a significant impact on how they perceive themselves and how disabled they are. The issue of access can limit the participation and integration of people with disabilities in society. "The shift from medical model to social model of disability" refers to the transition from an individual, medical point of view to a structural, social point of view. This shift, instead of treating impairment as something restricted to the individual's physical body, recognizes that disability is also a product of society. It is imperative to understand that disability is neither a purely medical nor a purely social issue, although the medical model and the social model are often interpreted as opposing views. As a result, PwDs face challenges that affect both their health and societal factors (WHO, 2011). The double vulnerability among the women with disabilities is depicted in Table 5.1.

As a result of social exclusion, people with disabilities cut themselves off from a wide range of activities, such as social, cultural, educational, and entertainment events. Even during holidays, family gatherings, and special occasions, their need to belong is even more pronounced, and their feeling of guilt and loneliness deepen. Their economic inferiority and segregation are maintained and reinforced by the fact that society is not willing to interact with them in social and cultural contexts. The study was conducted to find out the socio-cultural barriers that disabled people face. It looks at their social position, how society responds to them, and the barriers to their employment opportunities. The stark reality shows the existence of discrimination, poverty, and the absence of social respect, even though there are constitutional provisions for equal and special rights (Sharma, 2024).

The research of Gartrell et al. (2018) is one of the studies in this broader context, where it was supposed to inquire about a larger scheme of how socio-cultural beliefs are influential not just in attitudes and behaviors towards the disabled but also in the way policies and strategies can be designed to promote access and inclusion of the latter. Purposely, the field sites were chosen so that it would be possible to investigate the impact of culture on the disability experience and, more specifically, on that of women (Gartrell et al., 2018).

Malaita (patrilineal) and Isabel (matrilineal) were selected because the descent systems of these areas affect land buying, family relationships, and the distribution of responsibilities within a household—important socio-cultural factors that determine the existence and extent of disability, as well as the conditions for the sufferer. The Local Co-Researchers and Community-Based Rehabilitation (CBR) Officers used family and institutional networks to gain access to the communities. Individuals with disabilities were discovered through the CBR Disability Register of the Ministry of Health, and with the assistance of provincial CBR Officers. Fieldwork was done from November 2011 to June 2012 in Takwa, Malu'u, and several villages in Isabel, which resulted in 48 in-depth interviews with persons with disabilities and their households, as well as 37 key informant interviews (Gartrell et al., 2018).

All the interviews were recorded with informed consent, translated into English, and then manually coded through an inductive method that made certain themes arise straight from the narrations of the participants (Saldaña, 2021). The continuous work of the team reviewing has further made the interpretations more precise and has also reinforced the analysis (Taplin et al., 2002). The results reveal the extent to which the patrilineal and matrilineal systems influence access to land, inheritance, social support, and the expectations of different genders—this is also true for those factors that multiply the double vulnerability of women with disabilities. Ethical approvals were granted by Monash University and the Solomon Islands National Health Research and Ethics Committee, and to maintain confidentiality, pseudonyms were assigned to the participants.

Table 5.1 Dimensions of double vulnerability among women with disabilities

Dimension of vulnerability	Description	Key evidence
Gender-based discrimination	Women with disabilities are excluded from women's organisations, denied participation, and subjected to gender norms that restrict autonomy.	Exclusion from women's events; denial of reproductive choice; patriarchal norms reduce decision-making ability.
Disability-based stigma & cultural beliefs	Disability is often interpreted as a curses, taboo violations, or black magic; mothers blamed; disabled women were viewed as less valuable and incapable of fulfilling gender roles.	Solomon Islands studies show blame towards mothers, shame, exclusion; disability tied to notions of impurity or incompetence.
Land and inheritance inequalities	Patrilineal and shifting matrilineal systems restrict women's land access, worsening economic insecurity for disabled women.	Matrilineal → patrilineal shift; disabled women losing rights due to changing land systems; seen as outsiders in bilateral descent.
Economic marginalisation	Disabled women face the lowest employment rates, limited skills training, higher costs of living, and widespread employer bias.	ESCAP employment gaps; stigma from employers; inaccessible workplaces; limited vocational training.
Limited access to education	Girls with disabilities face stigma, unsafe environments, inaccessible schools, a lack of inclusive materials, and lower family prioritisation.	Clarke and Sawyer (2014); lack of ramps, assistive technology; families keeping disabled girls at home.
Weak social support networks	Kin-based support systems often fail disabled women; they are excluded from tool sharing, labour support, and community activities.	Kin networks bypass disabled women; exclusion from gardening, fishing tools, communal labour; reliance on fragile households.
Violence and legal capacity denial	Disabled women face heightened risks of gender-based violence, denial of legal autonomy, and guardianship restrictions.	Scully (2014), Arstein-Kerslake (2019); evidence of unpunished violence; denial of sexual and reproductive rights.
Disaster-related exclusion	In disasters, disabled women experience inaccessible evacuation centres, unsafe shelters, a lack of privacy, fear of abuse, and disproportionate injury.	Vanuatu studies: WASH failures, crowding, caregivers avoiding shelters, threefold injury risk in Cyclone Pam.
Menstrual stigma and WASH barriers	Menstrual health needs are ignored; inadequate sanitary supplies, lack of privacy, and water worsen exclusion during disasters.	Wilbur et al. (2022): inadequate pads; inability to dry reusable pads; unsafe WASH facilities; added caregiver burden.
Caregiving burdens and household poverty	Families (especially women caregivers) bear increased labour, lose income, and face long-term poverty linked to disasters.	Caregivers quitting jobs, increased incontinence during crises, and multigenerational impoverishment.
Limited access to SRH services	Disabled women are last to receive family planning, antenatal care, or STI treatment; stigma from health workers reduces access.	UNFPA & UNICEF evidence: providers assume disabled women are not sexually active; MISP is inconsistently implemented.
Policy invisibility and inadequate representation	CRPD, CEDAW, and Sendai commitments exist but are not implemented; disabled women are not represented in planning.	Policies are symbolic, not operational; women with disabilities are rarely included in DRR committees or gender programs.

Based on the above retrieved study data, various social determinants were studied that intensify disability-based disadvantages in the Solomon Islands, particularly for women experiencing compounded vulnerability. The multi-person socio-cultural systems that exist in the Solomon Islands, together with gender and land inheritance practices, create a situation of disadvantage for the entire disabled community in the country. Solomon Islanders' identity and social acceptance are intimately connected to land, and accessing it is largely determined by the matrilineal system in provinces such as Isabel and by the patrilineal system in the Malaita area (Otter, 2002). However, because of the combined effects of Christianity, colonialism, and being a part of a cash economy, which have given men the upper hand in land rights over women even in traditionally matrilineal areas, these traditional land rights systems are gradually disappearing (Maet al. &a, 2008; Peterson et al., 2012).

Changes in the land rights situation impact women with disabilities the most, as they are the most vulnerable of all. The women with disabilities' main source of income through the land rights that they were holding will not only cause them to lose their community participation but also put them in a long-term state of insecurity. Bilateral descent norms, though they seem quite adaptable, usually result in women and their children being considered "outsiders" and thus, they are denied access to the help and material resources that are available to the rest of the community (Burt, 1982; Hogbin, 1939; Keesing, 1982).

The power relations between genders further propagate these structural disadvantages. Women residing in patrilineal families are not allowed to partake in making decisions, have no or very limited freedom in choosing their partners, and have no or very little access to education and job opportunities (SINSO, 2009). The inequality is severe for women who are disabled since they suffer from being doubly marginalized—first, their being women and then their disability. On the other hand, the disabled women, when viewed through this lens of gender expectations, are burdened with three—gender, disability, and poverty—which often result in their being socially isolated and financially dependent on the fragile household arrangements (Gartrell et al., 2018). Studies indicate that women with disabilities undergo a greater degree of vulnerability when it comes to gender-based violence, social exclusion, and lack of support, and the culprits are not usually punished (Astbury & Walji, 2013; Spratt, 2013).

Cultural beliefs surrounding disability add to this double vulnerability. In various rural areas, disability has been associated, for a long time, with ancestral curses, violating taboos, or black magic, and the like, which usually result in blaming the mothers for their children's disabilities (Gartrell et al., 2018). These opinions lead to the imposition of shame, disregard, and cutting off both emotional and material support. Mothers of disabled children often face the difficulty of being in a situation where they have to contend alone, particularly in families that are already under the strain of poverty or insecure landholding. Moreover, kids with disabilities are very likely to undergo poor early inclusion, limited access to schooling, and reduced parental care, thus, factors that predispose them to long-term exclusion (SINSO, 2009). For female children, the disability-plus-gender-bias combo leads to even more precocious and harsh forms of discrimination.

Disability disadvantage is also influenced by social exclusion and weak support networks. In kin-based cultures where mutual assistance decides who gets what in terms of tools, land, and community resources, persons with disabilities—especially women—are almost always isolated from the support generally given to relatives. Even when they are economically active, stigma related to disability can stop them from being considered completely independent adults or rightful members of their kin groups (Gartrell et al., 2018). Disabled women are especially at risk of being deprived of such resources as gardening tools, fishing canoes, inheritance rights, or community labor support, all of which only serve to further their isolation.

Limited infrastructure is one of the main factors that prevent people with disabilities from participating in productive work, which is an essential factor for social recognition and contribution to the family in rural areas. The rough terrain, remote locations where the farms are, and the lack of proper facilities make it almost impossible for people with different abilities, or those who cannot see or hear, to work in the fields or go fishing. The negative cultural beliefs that people with disabilities "are not able to contribute anything" have pushed them out of work altogether, resulting in lower status within the community and increased poverty among the households (Gartrell et al., 2018). For women, who are already considered less valuable in society than men, as they are only seen as performing domestic work and taking care of others, the inability to do such work because of the disability leads to even more severe social isolation and economic insecurity.

5.3 Barriers and Structural Inequalities to Participation

Disability, in situations where gender norms, land ownership systems, and sociocultural beliefs intersect to form a compounding disadvantage, is crucial to understanding the experiences of women with disabilities. In the Solomon Islands, participation is not limited to persons with disabilities. However, it is also influenced by the larger structural forces of the patrilineal and matrilineal descent systems, insecure land rights, gendered division of labor, and specific stigma around disabilities. These forces limit access to education, livelihoods, mobility, and social interaction (Gartrell et al., 2018). Such barriers to participation are worsened by the cultural beliefs that view disability as the result of curses or violations of taboo, as well as by the consequent weakening of social support networks and appurtenant persistence of gender inequities (Maet al. &a, 2008; Spratt, 2013). Ultimately, the involvement of women with disabilities in the community is less a matter of personal ability but rather a question of the structural constraints imposed by the social, cultural, and economic systems.

5.3.1 Inequalities in Employment

Nevertheless, the extent of labor-market exclusion remains largely unknown; it is not explicitly evident in the employment statistics, presumably because this type of employment data is not disaggregated by disability. Due to the scarcity of job statistics broken down by disability, comparing the employment rates of those with disabilities to those without disabilities is complex and constrained. Furthermore, the disparities between internationally recognized employment statistics and disability-specific data raise doubts about the validity of some analyses. Due to both their gender and their condition, women with disabilities face double discrimination. In most Pacific nations, women with impairments have much lower employment rates than men with disabilities. The only country where women with impairments are more likely to work than men is Sri Lanka (ESCAP, 2015).

Data from the Pacific show that there are gender differences in the employment of persons with disabilities, and the situation is such that women are always in the lower category. In American Samoa, the difference in the employment rate of disabled men and women is about 3.6%, but the gap in Guam is bigger at about 11.7%, which means that women with disabilities are more excluded from the formal workforce. In the same way, both the Federated States of Micronesia and Samoa face the same problem, with employment gaps of 7.3% and 8.4%, respectively, signaling barriers that limit women's access to paid work. The Northern Mariana Islands have one of the most significant employment gaps in the region at 10.4%, indicating a significant gender-based disadvantage. The persistent employment gap in these Pacific contexts implies that the women working in the disability sector encounter margins, which are intersections of discrimination, through limited education, restricted mobility, cultural expectations of gender roles, and inadequate policy support. This whole situation makes their economic participation more difficult and increases their vulnerability (ESCAP, 2015).

In the Pacific, employment inequality for women with disabilities is influenced by a variety of factors—structural, social, and attitudinal—that hinder their ability to join the workforce. The whole region still lacks reliable disability-disaggregated statistics, which makes it impossible to make a very accurate comparison of the employment situation of disabled women, disabled men, and the non-disabled (ESCAP, 2015; WHO, 2011). Despite that, the data that is available show very clearly that there are gender gaps, as has been the case in many Pacific Island countries, such as Samoa, Micronesia, and Guam, where women with disabilities are often much less likely to get employed than men with disabilities. This is a result of the long-established gender norms, limited access to skill training, and fewer opportunities for financial independence.

Attitudes that are negative and discrimination as a practice remain the most substantial barriers that Pacific women with disabilities have to deal with. In this area, the employers are often biased and think that people with disabilities—especially women—are not able to do the work, are not worth the price, or are even going to be a burden on the company. Therefore, hiring rates are low, and workplace opportunities are limited. Such views are further strengthened by the stigmatization of society, which often relates a disability to shame, curse, or the family's bad luck. Consequently, the majority of disabled women are totally discouraged by their families and communities from trying to find jobs, which in turn, reduces their chance of being accepted in the economic sphere to almost nil (ESCAP, 2015; Groce, 2006; Ho-ting & Wong Ming-Wai, 2015).

The infrastructural challenges and the limited accessibility for disabled people are other issues that keep women with disabilities out of the labor market. In several Pacific Island locations, public transport is practically non-existent, and workplaces lack ramps or appropriate signs for people with disabilities, while communication facilities such as sign language interpreting or screen-reading assistive technology are not offered. These barriers to accessing the labor market hit women hardest because they are usually the ones who suffer from limited mobility due to their caregiving duties or because of the social stigma that considers them as housewives. Accessible environments, job information, and safe transportation are the main factors that enable women to participate in the workforce; without them, these opportunities become almost impossible to access (ESCAP, 2015).

Barriers in education and training further complicate the situation of inequalities. In the Pacific, women with disabilities are not only less likely to complete primary or secondary education compared to their male counterparts but also are not able to access any vocational training or higher education, thus resulting in no qualified positions available for them. Besides, those who have the chance to receive education or training usually do so before the disability starts, which is indicative of the lack of inclusive education systems and transition-to-work support. Moreover, limited social networks imply reduced exposure to job opportunities, and thus many women remain "invisible" in the labor market, where hiring is often influenced by community connections (Adioetomo et al., 2014; O'Keefe et al., 2007; WHO, 2011).

Legal and policy obstacles, in addition, make participation more difficult. In some nations, to the north and east of the continent, the law has not been changed and still necessitates that people with disabilities be "fully able-bodied" to work in certain occupations or hold public office, thus giving rise to the situation whereby persons with disabilities are excluded by default. Unyielding work setups, no reasonable accommodations provided, and the lack of discrimination laws make the situation even worse for them. Women who are disabled and get their disability later in life, primarily through work-related accidents, are more likely to be fired or get no support in their effort to return to work (Colbran, 2010; ESCAP, 2015).

Over the Pacific, women with disabilities suffer consequences of labor-market exclusion that are so deep that not only gender but also the disability-related discrimination is to blame for it. For example, ESCAP data imply that women with disabilities in the region are employed only one-fifth as often as men with disabilities, which is an indication of strong social norms that consider them to be less capable or less productive when it comes to formal work environments. Moreover, even if the Pacific women with disabilities are employed, they still will be in worse working conditions, be paid less, and have limited promotions, not only compared to their male colleagues with disabilities but also with other non-disabled women. Access to education and vocational training, which is restricted to them, is another factor contributing to their inability to secure skilled employment. Financially independent Pacific women with disabilities, through either earning or marriage, typically have been so before becoming disabled, which shows the powerful structural barriers preventing them from accessing decent work after acquiring an impairment. On the one hand, such trends are observed, while on the other hand, so far there has been little, however strong analysis due to the lack of disaggregated by sex and disability employment statistics across Pacific Island nations, which highlights the data gaps in the region that have not yet been resolved (ESCAP, 2012, 2015).

5.3.2 Inequalities in Education

Throughout the Pacific region, women with disabilities encounter numerous and overlapping obstacles that substantially limit their educational possibilities from the very early stages of development and throughout their life cycle. The majority of disabled children, particularly girls, are not able to enter school, as the families living in rural and low-income areas often do not send them to school, believing that education for such children is neither necessary nor worth the trouble. Stigma plays a huge role in this negative attitude as disabled people are often seen as disgraced, cursed by the gods, or not fit for society, since girls with disabilities are not considered able to learn or contribute in any way. Consequently, disabled girls have to stay at home doing household chores, or they are afraid of going out and getting hurt; thus, their academic and social growth is equally hampered (Clarke & Sawyer, 2014).

Poverty, infrastructure, and environmental limitations still restrict access, even in areas where education is provided. In the Pacific Islands, many schools do not have the type of buildings that are compatible with the accessibility standards, and the girls who have physical or sensory disabilities are unable to participate in school life, as there are no ramps, layout toilets, safe walkways, or inclusive classrooms. Moreover, the lack of easily accessible learning materials in braille, large print, assistive technology, or sign language support not only pushes girls with disabilities out of classroom activities but also pushes them to the back of the learning process, literally making them the last. The lack of teacher training in inclusive pedagogies has further kept these children isolated, as teachers often feel unprepared or are simply reluctant to meet the diverse educational needs of these children. The limited provision of accessible learning environments, such as the lack of sign-language interpreters, mobility-friendly classrooms, and adapted learning materials, further shrinks their access (Sharma et al., 2016). The deep-rooted stigma surrounding disability—especially in rural areas—often prevents girls with disabilities from being educated because families fear discrimination, bullying, or think that education is "wasted" on them. Economic constraints also hit harder on girls with disabilities, as families in financial difficulties usually choose to keep non-disabled children or boys over girls in school (ESCAP, 2015).

Socio-economic and geographic differences aggravate the challenges mentioned. The majority of Pacific communities are situated in regions that are hard to access, and, thus, transportation, which is both costly and complex, becomes a barrier to girls with disabilities attending school. Sometimes, even when special or inclusive schools are available, they are in such far-off places that mostly the girls will not be able to reach them, thus creating an accessibility gap that takes a toll on girls disproportionately. Another factor that weighs heavily is poverty: families that are financially struggling usually put the education of boys or non-disabled children first, and thus, girls with disabilities are left last for the very few educational resources. There are also the cultural standards that call for girls to always stay near home, together with the fear of violence and the dark side of our society, which, based on one's gender, further deters girls from attending school (Begum et al., 2019; Brown, 2024).

The obstacles that girls with disabilities go through are not just confined to the primary level; they face extreme limitations in the transition paths to secondary schooling, vocational training, and even the job market. They are hardly ever given the opportunity for career counseling, mentorship, or the community support necessary to continue their studies. The combination of structural inequalities in education systems and deeply rooted patriarchal norms sets up a persistent cycle wherein disabled women are still the majority of those excluded from formal education; thus, their socio-economic status is limited to that of the poorest and most marginalized, which in turn affects the whole community negatively in the Pacific due to the strengthening of the intergenerational disadvantage among the women (Thomas & Chandra, 2019; Thomas & Kumar, 2020).

5.4 Disasters and Women with Disability

Despite the greater global commitment to this approach, persons with disabilities are still commonly neglected, not listened to, and not taken into consideration in disaster risk reduction (DiDRR) processes and planning. This is especially evident in the case of women with disabilities, who suffer from multiple disadvantages, being that they are women, they have disabilities, and they are poor, and these are exacerbated during natural disasters. The disaster experience is challenging for women with disabilities, and they are also the ones with the least access to institutional support during the different phases of a disaster—preparedness, response, and recovery. Moreover, besides the threat of violence that comes with disasters, they are also more prone to suffering from other harms due to the escalating violence that comes with the disaster. Women with disabilities are forced to rely on the social capital of their homes and neighbors for support when there are no official resources available. They "recover" in whatever way they can, such as taking short-term loans, cutting back on food, or moving, all of which have a substantial negative impact on their long-term resilience (Gartrell et al., 2020).

The overlapping vulnerabilities that women with disabilities in the Pacific region suffer from during disasters are indeed very much amplified. In fact, in Vanuatu—one of the world's most disaster-prone countries—the combination of the factors gender, disability, and menstrual stigma imposes cumbersome restrictions on the participation of women and girls with intellectual disabilities in everyday life even before a crisis strikes (Baker et al., 2017; Wilbur et al., 2022a, 2022b, 2022c). Caregivers usually do not allow girls with intellectual disabilities to leave their homes, and the reasons for this are: the fear of sexual violence, community discrimination, and the embarrassment that may be caused by menstrual leakage or public removal of menstrual materials. Restrictions become even tighter during menstruation, resulting in a baseline of limited autonomy and social exclusion that is then exacerbated by disasters (Wilbur et al., 2022a, 2022b, 2022c).

Evacuation environments turn especially dangerous and inaccessible for women with intellectual disabilities during emergencies like Tropical Cyclone Harold and the Ambae volcanic eruption. Research indicates that during Cyclone Pam, people with disabilities were almost three times more prone to get injured than others, and women with disabilities had notably less access to sanitation facilities than their male counterparts, which is a clear indication of gender disparity in disaster impact (Baker et al., 2017). Caregivers said they were avoiding evacuation centers because of overcrowding, lack of privacy, and fear of abuse; thus, some families chose to stay in damaged or unsafe houses instead of risking being in communal shelters that are insecure (Baker et al., 2017; Downing et al., 2021). Such choices demonstrate the structural exclusion that is inherent in disaster response systems, where the evacuation facilities do not cater to the privacy or safety requirements of women who menstruate.

Economic uncertainty makes the situation even worse. Caregivers who have been carrying full-time caregiving responsibilities often lose the opportunity to earn income or have their income reduced during and after disasters (Wilbur et al., 2021). The trauma and stress that come along with cyclones make the situation worse for the young ones with intellectual disabilities and their caregivers because the former may become incontinent or display more difficult behaviors, which again leads to increased caregiver workload and makes it impossible to maintain livelihoods. Photovoice accounts describe caregivers who quit their jobs to be with the young woman at all times, making not only the young woman but also entire multigenerational households poorer with each disaster (Downing et al., 2021; Wilbur et al., 2021). This cycle shows how disasters promote the structural poverty of households with disabilities.

Menstrual health necessities highlight the inadequacy of current humanitarian interventions once more. Caregivers who received menstrual pads and reusable materials in the emergency hygiene kits after Cyclone Harold, however, reported regularly that the amount and kind of menstrual products were not enough for the individuals with disabilities, mainly those with incontinence (Baker et al., 2017; Wilbur et al., 2022a, 2022b, 2022c). There was no way to dry the reusable pads during the days-long rain; using disposables all the time was not an option because it was too expensive. The key informants noted that giving each family one pack of pads was not enough, and they were right, as families with disabled members would need larger quantities and various types of products, such as adult diapers (Wilbur et al., 2022a, 2022b, 2022c).

Safe and private WASH facilities are essential for menstrual health, yet WASH failures are common after natural disasters. Vanuatu case studies show that the lack of water after cyclones put tremendous strain on caregivers who had to pay for truck rides to the sea to bathe their daughters—a burden that already low-income families found impossible to bear (Wilbur et al., 2021). These service gaps underscore the need to prioritize disability-inclusive WASH planning in national disaster preparedness initiatives.

Lastly, the article (Wilbur et al., 2022a, 2022b, 2022c) emphasizes that menstrual health interventions should start long before disasters strike. Humanitarian workers indicated that raising awareness, menstrual health education, and planning that includes people with disabilities are rarely done in times of peace, thereby creating gaps that cannot be filled during quick emergency responses (Wilbur et al., 2021). If no preventive measures are taken to involve disability organizations and families, women with cognitive impairments will not be considered during humanitarian planning, and thus, they will be excluded again in every disaster cycle. The research strongly suggests that menstrual health, disability inclusion, and gender protection should be treated as issues that cut across the whole disaster risk reduction and response process rather than only during post-disaster relief (Handbook, 2011; Sommer, 2012; Sommer et al., 2016, 2017; Wilbur et al., 2022a, 2022b, 2022c).

From our analysis, it has been concluded that catastrophes aggravate the disparities that women with disabilities face, particularly in the Pacific region. Notably, these women are caught up in the web of the combined effects of gender discrimination, disability-based exclusion, menstrual stigma, and impoverishment to a great extent. As a result, their safety, movement, and access to help during an emergency are minimal. The case studies conducted in Vanuatu provide evidence of the problem's magnitude by demonstrating that the women with disabilities, along with their caregivers, have to go through unsafe and long-term suffering situations due to the lack of accessible evacuation centers, inadequate WASH facilities, a shortage of menstrual supplies, and fear of violence. Our analysis indicates that increased caregiving burden, loss of income, and lack of inclusive humanitarian planning continue to be the reasons that trap these households in structural poverty. The recurring patterns justify advocating for the enactment of disability-inclusive disaster risk reduction measures that involve DPOs, caregivers, and women with disabilities in the cycle of disaster preparedness, response, and recovery. Based on the evidence, we insist that the priority be given to the creation of inclusive WASH systems, proper menstrual health planning, and safe evacuation environments long before disasters occur to break the exclusion cycle that women with disabilities go through in every disaster situation.

5.5 Reproductive and Health Service Access Among Women with Disability

The use of sexual and reproductive health (SRH) information and services by young people in Pacific Island Countries and Territories (PICTs) is still very low, although SRH services are available (Harrington et al., 2021). In the world, young adulthood is a crucial period of development, marked by both opportunities and difficulties. One difficulty is the lack of access to healthcare services. For instance, the Society of Adolescent Health and Medicine, based on evidence from more than 40 publications in the United States, concluded that young adults aged 18–25 years experience the highest rates of mortality and unintended pregnancies, while also having the lowest access to healthcare compared with both younger (10–17 years) and older (26–30 years) age groups (Walker-Harding et al., 2017). Besides, considering the uniqueness of the sexual health needs, young adults are required to have precise information and quick access to contraceptives to avoid the negative effects of unintended pregnancies, abortion, childbirth, and untreated sexually transmitted infections (STIs) (Baigry et al., 2023; Prata & Weidert, 2020; WHO, 2020, 2022).

Access to sexual and reproductive health (SRH) information and services in the Pacific region is influenced by a mix of factors, including culture, the economy, and geographic location (Baigry et al., 2023). A study conducted in Papua New Guinea to get a better picture of the distribution and utilization of public funds highlighted the following reasons: shortage of healthcare professionals, funding delays or non-payment to healthcare providers, and limited availability of drugs (Howes et al., 2014), which in turn negatively impacted the delivery and accessibility of services. Rural and remote area residents and people with particular needs, like youth, disabled persons, and HIV-positive individuals, along with those identified as part of the LGBTQI community, are most likely to encounter difficulties in accessing SRH services (Baigry et al., 2023). Thus, the global efforts to guarantee the right to SRH and quality and accurate SRH information and services are further weakened (Sciortino, 2020), leading to situations like low contraceptive prevalence rate and high unmet need for contraceptives among young sexually active women in some PICTs (UNPF, 2013).

SRH rights are explicitly stated in the Minimum Initial Service Package (MISP) for reproductive health in emergencies, and its requirements include the provision of services at the earliest possible time at the onset of disasters. The situation, however, is different in the Pacific region, where assessments indicate that the MISP standards are inconsistently followed and that few adaptations are explicitly made for people with disabilities (Lisam, 2014).

In the case of Fiji, Vanuatu, Guam, and Tonga, the humanitarian assessments after the major cyclones have always pointed out that women with disabilities are the last to receive family planning services, antenatal care, postnatal care, and interventions for pregnancy-related emergencies. Besides, social stigma, which is the outcome of gender-related and cultural beliefs, often causes health workers to reject attitudes who might mistakenly think that women with disabilities are not sexually active or do not need SRH services. These incorrect notions form a cycle of invisibility that increases the risk during crises. Therefore, the inclusion of persons with disabilities in SRH programming, though crucial, is often neglected in disaster preparedness in the Pacific (ESCAP, 2023; UNICEF, 2017).

5.6 Support Needs for Women with Disability

Pacific women with disabilities endure the worst possible scenario, characterized by gender inequality, stigmatization of disabilities, and isolation in the island context with minimal services now and then. In most of the Pacific, the care for the family, especially women with disabilities, has to be done by the family mostly, while the formal support structures in health, education, safety, and livelihoods are already very few or too far to reach. Cultural norms, limited mobility, poverty, and weak policy frameworks are among the structural barriers preventing these women from enjoying the rights that come with their participation and protection. Therefore, it is necessary to understand the support requirements of female disabled persons in the Pacific by ascertaining how social, cultural, and institutional factors relate to their marginalization, thereby making the need for inclusive policies, strong community-based support, and rights-centered interventions across the region apparent and even more urgent.

The upcoming measures in the Solomon Islands need to place the support of gender equality in the landownership and descent systems first, because these structural arrangements are still working against women, including disabled women. The fight against gendered power relations is critical since the changes in descent systems, patriarchal norms, and external forces have all contributed to the weakening of women's power and the increase of their vulnerability. Policies should focus on creating local livelihood opportunities, which will enhance disabled persons' ability to participate productively in community and household economies. Access to education, vocational training, and skill-building initiatives is crucial to women with disabilities who are facing multiple overlapping barriers (Gartrell et al., 2018). The rights-based interventions needed to strengthen gender- and disability-inclusive disaster risk reduction are depicted in Fig. 5.1.

The economic security, self-esteem, and social inclusion would all be increased by these activities. Strengthening locally rooted support systems is an additional key future priority. Community-based structures can ensure social and material support flows consistently to families with disabled persons, especially when women are the primary caregivers. These methods would reduce isolation, alleviate hardship, and enable individuals to participate in daily life with much greater quality. A social determinants and rights-based framework, compatible with the UN CRPD principles, will be an excellent basis for future policy and program development in the Solomon Islands. This method acknowledges that disability disadvantage cannot be solved

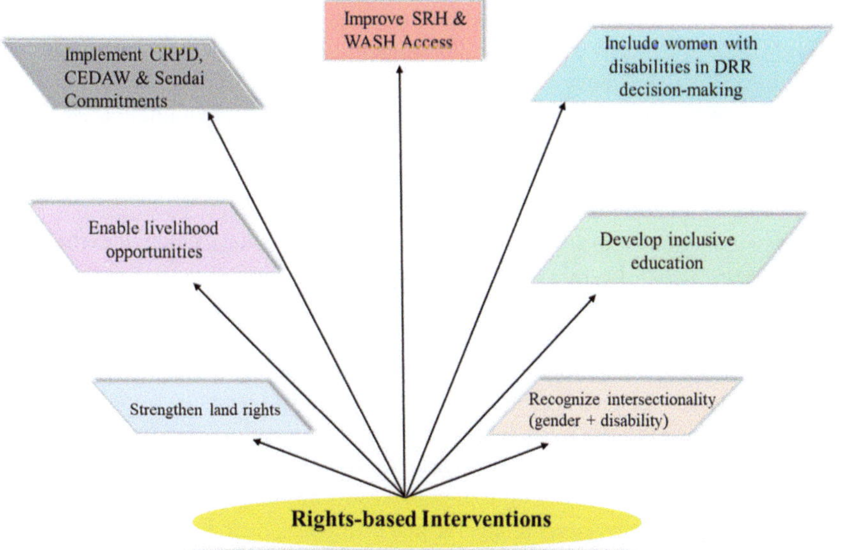

Fig. 5.1 Key rights-based interventions needed to strengthen gender- and disability-inclusive disaster risk reduction. This figure highlights a set of rights-based interventions essential for advancing inclusive disaster risk reduction for women with disabilities. These interventions include implementing international commitments such as CRPD, CEDAW, and the Sendai Framework; improving access to sexual and reproductive health (SRH) and WASH services; and ensuring the participation of women with disabilities in decision-making processes related to disaster planning and response. Additional priorities involve expanding livelihood opportunities, developing inclusive education systems, strengthening land rights, and recognizing intersectionality between gender and disability. Collectively, these measures contribute to a more equitable and rights-based approach to disaster resilience

without addressing gender inequalities, which are the root of the problem, thus positioning women with disabilities at the heart of any long-term inclusive development strategy (Gartrell et al., 2018).

Vanuatu women with disabilities are among the least supported persons, owing to the shortage of services, deeply entrenched cultural barriers, and the fall of the traditional family support system. Close relatives—primarily women, who provide the majority of care—so there is an urgent need for organized systems that reduce dependence and promote independence. The services that should be provided for disabled women in the Pacific region include counseling, rehabilitation, educational assistance, and job placement, but these are either not offered at all or are offered very little. There is a need to establish robust community rehabilitation programs and to integrate people with disabilities into the formulation of health, protection, and social participation policies to address gaps in these areas. Women living with disabilities will need to have statutory and institutional protections that are very strong so that their rights are not violated, especially in situations where their safety is threatened, marriage is hard to enter, or education is unavailable to them. Such a situation calls for implementing legislative reforms, backed by campaigns to change cultural attitudes towards women's roles and abilities.

Additionally, they would require leadership training, skills development, and the running of strong women's units in disability organizations to facilitate their meaningful participation as opportunities for capacity-building. Accurate data broken down by gender, disability, and age is essential for developing effective policies and tackling structural inequality. The cooperation between governments, non-governmental organizations, communities, and regional bodies, along with the introduction of dedicated disability strategies and the provision of sufficient resources, is imperative for the success of coordinated support that will empower women with disabilities to utilize their abilities to the fullest.

WWDPN has the potential to become a dynamic and influential association of disabled women in the Pacific if it receives the support and financial backing it needs. It expresses a desire to work collaboratively with organizations that share an interest in disability and women's issues. WWDPN, a new organization, does not view itself as taking over the role of existing regional organizations, but rather it believes that women with disabilities need a special place to learn their way, to develop, and confer with each other on their concerns. The WWDPN, as a coalition of women with disabilities, is well-positioned to represent and advocate for the issues affecting disabled women. While some women's groups have included disabled women in their programs, the struggle to change the perception of the women's movement and the national women's machinery remains to ensure that the issues of women with disabilities are part of the national plan of action. Identifying the barriers faced by women with disabilities in the Pacific is the first step WWDPN will take. Among these actions are cooperation with local, regional, and international organizations, the formation of a national network of disabled women, and raising community awareness of the issues facing women with disabilities.

5.7 Conclusion

Women with disabilities in the Pacific islands are indeed suffering a lot. They experience discrimination not only as women but also as disabled individuals. Additionally, these women are primarily found in areas most affected by both poverty and nature's wrath. The situation gets even worse due to factors such as cultural practices, uneven land distribution, and a shortage of education, job opportunities, and community support. One of the major factors keeping their already fragile situation from deteriorating is natural disasters; this is particularly true during shelter evacuations, WASH facilities usage, and when their requirements are not taken into account in the assistance given to them. The chapter has illustrated that the disability-related disadvantage is primarily connected with the gendered power relations, which are at the very heart of the Pacific societies. A Social Determinants and Rights-based Framework will be crucial in future policy and programming support, which will entail legislative reforms, inclusive livelihood initiatives, accessible SRH services, and strong community-based support systems. In addition, it will mean an upgrade in data gathering, giving power to women with disabilities to take part in decision-making, and working together of governments, NGOs, DPOs, and regional organizations to realize the dream of inclusive and sturdy communities throughout the Pacific. Ultimately, the transformation will be based on the official recognition of women with disabilities as bearers of rights and on their involvement in development processes related to equity and sustainability.

References

Adioetomo, S. M., Mont, D., & Irwanto, I. (2014). *Persons with disabilities in Indonesia: Empirical facts and implications for social protection policies*. TNP2K.

Arstein-Kerslake, A. (2019). Gendered denials: Vulnerability created by barriers to legal capacity for women and disabled women. *International Journal of Law and Psychiatry, 66*, Article 101501. https://doi.org/10.1016/j.ijlp.2019.101501

Astbury, J., & Walji, F. (2013). *Triple Jeopardy: Gender-based violence and human rights violations experienced by women with disabilities in Cambodia*. AUSAID. https://iwda.org.au/assets/files/20130204_TripleJeopardyReport.pdf

Baigry, M. I., Ray, R., Lindsay, D., Kelly-Hanku, A., & Redman-MacLaren, M. (2023). Barriers and enablers to young people accessing sexual and reproductive health services in Pacific Island Countries and Territories: A scoping review. *PLoS ONE, 18*(1), Article e0280667. https://doi.org/10.1371/journal.pone.0280667

Baker, K., Brown, T., Caleb, N., Iakavai, J., Marella, M., Morris, K., ... & Pryor, W. (2017). Disability inclusion in disaster risk reduction: Experiences of people with disabilities in vanuatu during and after tropical cyclone pam and recommendations for humanitarian agencies. Nossal Institute for Global Health, CBM Australia, Oxfam in Vanuatu, Vanuatu Society of People with Disabilities, Vanuatu Disability Promotion and Advocacy Association, Ministry of Justice and Community Services, Vanuatu National Disaster Risk Management Office: Melbourne, Australia.baker et al.

Begum, H. A., Perveen, R., Chakma, E., Dewan, L., Afroze, R. S., & Tangen, D. (2019). The challenges of geographical inclusive education in rural Bangladesh. *International Journal of Inclusive Education, 23*(1), 7–22. https://doi.org/10.1080/13603116.2018.1514729

Boyle, M. (2003). The dangers of vulnerability. *Clinical Psychology, 24*(4), 27–30.

Brown, C. (2024). *Equity and inclusion in Pacific education*. https://unesdoc.unesco.org/ark:/48223/pf0000391476

Burt, B. (1982). Kastom, Christianity and the First Ancestor of the Kwara'ae of Malaita. *Mankind, 13*(4), 374–399. https://doi.org/10.1111/j.1835-9310.1982.tb01001.x

Butler, J., Gambetti, Z., & Sabsay, L. (Eds.). (2016). *Vulnerability in resistance*. Duke University Press. https://doi.org/10.1215/9780822373490

Clarke, D., & Sawyer, J. (2014). *Girls, disabilities and school education in the East Asia Pacific region*. United Nations Girl's Education Initiative.

Colbran, N. (2010). *Access to justice persons with disabilites in Indonesia*. Background Assessment Report, AusAID. https://www.dfat.gov.au/sites/default/files/access-justice-2010.pdf

Downing, S. G., Benjimen, S., Natoli, L., & Bell, V. (2021). Menstrual hygiene management in disasters. *Waterlines, 40*(3), 144–159.

Economic and Social Commission for Asia and the Pacific. (2023). Asia-Pacific Population and Development Report 2023. In *Seventh Asian and Pacific Population Conference*. https://www.unescap.org/kp/2023/asia-pacific-population-and-development-report-2023

ESCAP, U. (2012). *Disability, livelihood and poverty in Asia and the Pacific: An executive summary of research findings*. https://www.unescap.org/sites/default/files/SDD_PUB_Disability-Livelihood.pdf

ESCAP, U. (2015). *Disability at a glance 2015: Strengthening employment prospects for persons with disabilities in Asia and the Pacific*. https://www.unescap.org/publications/disability-glance-2015-strengthening-employment-prospects-persons-disabilities-asia-0

Gartrell, A., Calgaro, E., Goddard, G., & Saorath, N. (2020). Disaster experiences of women with disabilities: Barriers and opportunities for disability inclusive disaster risk reduction in Cambodia. *Global Environmental Change, 64*, Article 102134. https://doi.org/10.1016/j.gloenvcha.2020.102134

Gartrell, A., Jennaway, M., Manderson, L., Fangalasuu, J., & Dolaiano, S. (2018). Social determinants of disability-based disadvantage in Solomon islands. *Health Promotion International, 33*(2), 250–260. https://doi.org/10.1093/heapro/daw071

Groce, N. (2006). Cultural beliefs and practices that influence the type and nature of data collected on individuals with disability through national census. In *International views on disability measures: Moving toward comparative measurement* (Vol. 4, pp. 41–54). Emerald Group Publishing Limited. https://doi.org/10.1016/S1479-3547(05)04004-2

Guilbert, J. J. (1999). The World Health Report 1998—Life in the 21st century. A vision for all. *Education for Health, 12*(3), 391.

Handbook, S. (2011). *Humanitarian charter and minimum standards in humanitarian response*. International Federation of Red Cross and Red Crescent Societies.

Hans, A., & Patri, A. (2003). *Women, disability and identity*.

Harrington, R. B., Harvey, N., Larkins, S., & Redman-MacLaren, M. (2021). Family planning in Pacific island countries and territories (PICTs): A scoping review. *PLoS ONE, 16*(8), Article e0255080. https://doi.org/10.1371/journal.pone.0255080

Hogbin, H. I. (1939). *Experiments in civilization: The effects of European culture on a native community of the Solomon Islands*. https://onesearch.slq.qld.gov.au/permalink/61SLQ_INST/1dejkfd/alma997245314702061

Hollander, J. A. (2002). Resisting vulnerability: The social reconstruction of gender in interaction. *Social Problems, 49*(4), 474–496. https://doi.org/10.1525/sp.2002.49.4.474

Ho-ting, C. M., & Wong Ming-Wai, F. (2015). *Baseline survey on employers' attitudes towards employment of people with disabilities*.

Howes, S., Mako, A. A., Swan, A., Walton, G., Webster, T., & Wiltshire, C. (2014). *A lost decade? Service delivery and reforms in Papua New Guinea 2002–2012*. NRI-ANU Promoting Effective Public Expenditure (PEPE) Project. https://www.pngnri.org/images/Publications/OP_-_201410_-_Howes_-_Lost_Decade_2.pdf

Keesing, R. M. (1982). *Kwaio religion: The living and the dead in a Solomon Island society*. Columbia University Press.

Lajoie, A. (2018). *Challenging assumptions of vulnerability: The significance of gender in the work, lives and identities of women human rights defenders* (Doctoral dissertation). https://doi.org/10.13025/16742

Lisam, S. (2014). Minimum initial service package (MISP) for sexual and reproductive health in disasters. *Journal of Evidence-Based Medicine, 7*(4), 245–248. https://doi.org/10.1111/jebm.12130

Lowe, H., Utumapu, M. A. F. A., Tevaga, P., Ene, P. I., & Mannell, J. (2025). Disability and intimate partner violence experience among women in rural Samoa: A cross-sectional analysis. *Disability and Health Journal, 18*(2), Article 101735. https://doi.org/10.1016/j.dhjo.2024.101735

Luna, F. (2009). Elucidating the concept of vulnerability: Layers not labels. *IJFAB: International Journal of Feminist Approaches to Bioethics, 2*(1), 121–139. https://doi.org/10.3138/ijfab.2.1.121

Maetala, R. (2008). Matrilineal land tenure systems in Solomon Islands: the cases of Guadalcanal, Makira and Isabel Provinces. *Land and women: The matrilineal factor*, 35–72.

Maya Dhungana, B. (2006). The lives of disabled women in Nepal: Vulnerability without support. *Disability & Society, 21*(2), 133–146. https://doi.org/10.1080/09687590500498051

Moodley, J., & Graham, L. (2015). The importance of intersectionality in disability and gender studies. *Agenda, 29*(2), 24–33. https://doi.org/10.1080/10130950.2015.1041802

O'Keefe, P. (2007). *People with disabilities in India: From commitments to outcomes.* Human Development Unit, South East Asia Region, The World Bank, 157, 295583-1220435937125. https://documents.worldbank.org/en/publication/documents-reports/documentdetail/358151468268839622

Otter, M. (2002). *Solomon islands human development report: Building a Nation.* Windsor, Government of the Solomon Islands/UNDP.

Peterson, N., Hamilton, R., Pita, J., Atu, W., & James, R. (2012, August). *Ridges to reefs conservation plan.*

Prata, N., & Weidert, K. (2020). Adolescent sexual and reproductive health. In *Oxford research encyclopedia of global public health.* https://doi.org/10.1093/acrefore/9780190632366.013.206

Saldaña, J. (2021). *The coding manual for qualitative researchers.* https://doi.org/10.29333/ajqr/12085

Sands, T. (2005). A voice of our own: Advocacy by women with disability in Australia and the Pacific. *Gender & Development, 13*(3), 51–62. https://doi.org/10.1080/13552070512331332297

Sciortino, R. (2020). Sexual and reproductive health and rights for all in Southeast Asia: More than SDGs aspirations. *Culture, Health & Sexuality, 22*(7), 744–761. https://doi.org/10.1080/13691058.2020.1718213

Scully, J. L. (2014). Disability and vulnerability: On bodies, dependence, and power. *Vulnerability: New essays in ethics and feminist philosophy*, 204, 209–210. https://doi.org/10.1093/acprof:oso/9780199316649.003.0009

Sharma, A. (2024). Socio-cultural challenges faced by people with physical disabilities. *Shahid Kirti Multidisciplinary Journal, 2*(2), 81–90. https://doi.org/10.3126/skmj.v2i2.62504

Sharma, U., Forlin, C., Sprunt, B., & Merumeru, L. (2016). Identifying disability-inclusive indicators currently employed to monitor and evaluate education in the Pacific island countries. *Cogent Education, 3*(1), 1170754. https://doi.org/10.1080/2331186X.2016.1170754

Solomon Islands National Statistics Office, Secretariat of the Pacific Community, & Macro International Inc. (2009). *Solomon islands demographic and health survey 2006–2007.* https://statistics.gov.sb/dhs-2006-2007/

Sommer, M. (2012). Menstrual hygiene management in humanitarian emergencies: Gaps and recommendations. *Waterlines,* 83–104. https://doi.org/10.3362/1756-3488.2012.008

Sommer, M., Schmitt, M. L., Clatworthy, D., Bramucci, G., Wheeler, E., & Ratnayake, R. (2016). What is the scope for addressing menstrual hygiene management in complex humanitarian emergencies? *A Global Review. Waterlines, 245–264,*. https://doi.org/10.3362/1756-3488.2016.024

Sommer, M., Schmitt, M., & Clatworthy, D. (2017). *A toolkit for integrating Menstrual Hygiene Management (MHM) into humanitarian response.* Columbia University, Mailman School of Public Health and International Rescue Committee. https://hdl.handle.net/10776/13247

Spratt, J., & United Nations Fund for Population Activities. (2013). *A deeper silence: The unheard experiences of women with disabilities: sexual and reproductive health and violence against women in Kiribati, Solomon Islands and Tonga.* https://pacific.unfpa.org/sites/default/files/pub-pdf/UNFPAReport-ADeeperSilenceA4PublicationLR3%283%29.pdf

Taplin, D. H., Scheld, S., & Low, S. M. (2002). Rapid ethnographic assessment in urban parks: A case study of Independence National Historical Park. *Human Organization, 61*(1), 80–93. https://doi.org/10.17730/humo.61.1.6ayvl8t0aekf8vmy

Thomas, F. B., & Chandra, S. (2019). The Challenges faced by women with disabilities in education. *Journal of Emerging Technologies and Innovative Research, 6*(4), 124–130.

Thomas, F. B., & Kumar, S. L. (2020). The challenges and barriers encountered by educators in managing students with intellectual disabilities (SWIDS) in inclusive classroom settings. *Open Access International Journal of Science & Engineering, 5*(2), 16–21.

UNICEF. (2017). *Situation analysis of children in the Pacific Island Countries.* https://www.unicef.org/pacificislands/media/661/file/Situation-Analysis-Pacific-Island-Countries.pdf

United Nations Population Fund. (2013). *I am not a lost cause! Young women's empowerment and teenage pregnancy in the Pacific.*

Walker-Harding, L. R., Christie, D., Joffe, A., Lau, J. S., & Neinstein, L. (2017). Young adult health and well-being: a position statement of the society for adolescent health and medicine. *Journal of Adolescent Health.*

Wilbur, J., Clemens, F., Sweet, E., Banks, L. M., & Morrison, C. (2022a). The inclusion of disability within efforts to address menstrual health during humanitarian emergencies: A systematized review. *Frontiers in Water, 4,* Article 983789. https://doi.org/10.3389/frwa.2022.983789

Wilbur, J., Morrison, C., Bambery, L., Tanguay, J., Baker, S., Sheppard, P., ... & Mactaggart, I. (2021). "I'm scared to talk about it": Exploring experiences of incontinence for people with and without disabilities in Vanuatu, using mixed methods. *The Lancet Regional Health–Western Pacific, 14.*

Wilbur, J., Morrison, C., Iakavai, J., Shem, J., Poilapa, R., Bambery, L., ... & Mactaggart, I. (2022). "The weather is not good": exploring the menstrual health experiences of menstruators with and without disabilities in Vanuatu. The Lancet Regional Health–Western Pacific, 18. https://doi.org/10.1016/j.lanwpc.2021.100325

Wilbur, J., Poilapa, R., & Morrison, C. (2022c). Menstrual health experiences of people with intellectual disabilities and their caregivers during Vanuatu's humanitarian responses: A qualitative study. *International Journal of Environmental Research and Public Health, 19*(21), 14540. https://doi.org/10.3390/ijerph192114540

World Health Organization, & World Health Organization. (2020). *Adolescent pregnancy. 2020.* https://www.who.int/news-room/fact-sheets/detail/adolescent-pregnancy

World Health Organization. (2011). World report on disability. In *World report on disability* (pp. 24–24). https://iris.who.int/handle/10665/44575

World Health Organization. (2022). *Report on global sexually transmitted infection surveillance, 2018.* World Health Organization. https://www.who.int/publications/i/item/9789241565691

Chapter 6
Indigenous Knowledge, Disability, and Ecological Wisdom

Abstract This chapter investigates the intricate interaction among the traditionally held knowledge systems of the Indigenous people, disabilities, and ecological wisdom in Oceania, while at the same time pointing out how the factors of colonialism, Western epistemology, and structural marginalization have affected the survival of the Indigenous PwD. It brings together interdisciplinary research, lived experiences, and regional case studies from the Pacific countries to criticize the main Western disability frameworks that ignore the Indigenous worldviews, which are based on interdependence, spirituality, relational identity, and connection to the land of the ancestors. It also points out that the Indigenous people's view of disability—that is, based on cultural belonging, social roles, kinship networks, and holistic well-being—is in sharp contrast with the biomedical and deficit models that have been imposed through colonial education, law, and medicine. The chapter also looks at the influence of climate variation on the Indigenous communities with disabilities and the risk of disasters that are heightened by the lack of traditional ecological knowledge, community-based support systems, and culturally grounded resilience strategies. Climate change and ecological loss, if not handled properly, would not only impede cultural continuity but also make Indigenous-led knowledge and governance structures, as well as disability-inclusive resilience systems, unavoidable pathways for rights, self-determination, and sustainable futures in the Pacific. The chapter ends with a strong call to decouple disability from the colonial past and the Western way of viewing things, by issuing a call to the Indigenous ways of knowing as the leaders of equitable and culturally sensitive responses to disability and disaster vulnerability in Oceania.

Keywords Indigenous knowledge · Oceania · Disability · Intersectionality · Resilience

S. M. Thomas and R. Veerabathiran, *Disability, Disaster, and Resilience in Oceania*, SpringerBriefs in Modern Perspectives on Disability Research,
https://doi.org/10.1007/978-981-95-8535-9_6

6.1 Introduction

Perhaps the most widely accepted definition of disability is a social phenomenon that affects people with limitations that prevent them from fully participating in society (Oliver, 2017). The UN General Assembly views disability as "an evolving concept that results from the interaction between persons with impairments and attitudinal and environmental barriers" (United Nations General Assembly, 2007). These relational conceptions of disability, which view disability as a social experience that exists inside our society's culture rather than within the individual, are supported by many disability experts (Goodley, 2024; Oliver, 2017). Therefore, other types of prejudice, like racism, classism, and sexism, exacerbate disability (Annamma et al., 2018; Goodley, 2017). In the past, disability may have been a positive trait for certain Indigenous societies. However, in the framework of settler-colonialism, disability is just one of several elements that undermine Indigenous self-determination and futurity (Lovern & Locust, 2013).

Indigenous PwD are frequently thought of as "doubly disadvantaged," since their participation in society is severely restricted due to both racial and disability discrimination. Due to their geographic location, financial difficulties, systemic racism, and a variety of social circumstances that may exacerbate their health, Indigenous PwD encounter obstacles when trying to access support services. According to social determinants of health models, Indigenous people are more likely than non-Indigenous people to suffer from chronic illness, injury, suicide, and death (Ineese-Nash, 2020).

According to the International Work Group for Indigenous Affairs, there are between 200 and 370 million indigenous people worldwide. Indigenous peoples live in every part of the planet. The International Labour Organization's Convention No. 169 on Indigenous and Tribal Peoples, adopted in 1989, and the United Nations General Assembly's Declaration on the Rights of Indigenous Peoples, adopted on September 13, 2007, are essential international agreements that define the rights of indigenous peoples. Indigenous peoples are defined by the ILO Convention No. 169 as those whose ancestors resided in the region prior to settlement or the creation of current state borders. Furthermore, the treaty stipulates that indigenous peoples have preserved their own social, economic, cultural, and political institutions in whole or in part. The ILO No. 169 Convention requires nations to take exceptional steps to conserve indigenous cultures, languages, and habitats, among other things, and acknowledges the unique rights of indigenous peoples to their historic residence locations and natural resources. Nevertheless, the agreement does not adopt a position regarding the definition of indigenous people (Joona, 2012; Sarivaara et al., 2013).

The intersectionality of Indigeneity and disability significantly points to a complex layering of marginalization through the processes of colonialism, racialization, and the imposition of Western knowledge systems that are not cognizant of the Indigenous presence. In the eyes of some Indigenous disability scholars, Western science and quantitatively based frameworks have long been the ones to define disability and control the discourse around it without any input from the Indigenous side, thereby building stories that are far from the truth. Norms of the "normalcy," which were imposed from outside, carried by colonial ideologies of civilized versus uncivilized bodies, formed a hierarchy where the Indigenous peoples were deemed inferior and, at the same time, the ones requiring diagnostic labels to be recognized and understood by the settler institutions. This dual marginalization—derived from race and ability—enlightens how colonialism and ableism come together to create a pathology of Indigenous existence (Gilroy et al., 2021).

Residential schools—their worldview—were one of the colonial institutions that definitely worsened this intersectional harm by classifying the Indigenous children as "disabled" if they could not cope with the systems that were purposefully suppressing the children's languages, learning styles, and depriving them of their cultural teachings. Trauma, cultural disruption, and forced assimilation were all interpreted by the non-Indigenous teachers and psychologists as the individuals' intrinsic impairment. Consequently, disability classification was turning into an effective social control mechanism that ensnared the Indigenous people in the net of dependency while at the same time making their dependence unattractive. Furthermore, such a situation was fostering the stereotypes that the indigenous population is uneducable or inherently deficient. These categories also resulted in the proliferation of intergenerational trauma and the psychological repercussions enduring over a sustained period of time, thus revealing the synergy of disability labeling and colonial oppression as the great restrictors of opportunities (Gilroy et al., 2021).

Moreover, the intersectional experience of Indigenous people, as highlighted in their social and cultural matters, is now reflected in global policy frameworks like the International Classification of Functioning (ICF), which ignore the input of Indigenous peoples in their development. The ICF and its application to Indigenous communities are inadequate because they fail to provide insights into the Indigenous ways of knowing and being regarding ability, relationships and support that are culturally tied henceforth, the combination of Indigenous status and disability issue is not only a question of human or mental difference but also a mirror of historical power factor, epistemic suppression, and continual fight for self-determination. Therefore, acknowledging these crossing forces is of paramount importance for promoting Indigenous-led, culturally based disability scholarship and for the reclamation of the authority of Indigenous knowledge systems (Senior, 2000; Toni, 2007; WHO, 2002).

The above-explained arguments together express the opinion that disability in the context of Indigenous peoples has been coextensive with the colonial oppression, the consequent epistemic elimination, and the discriminatory practices that characterized the entire history of the Indigenous peoples. It is the constant Westernization through different means like education, medical and legal frameworks, or even global policies that play a Major role in the way indigenous people's disabilities are acknowledged, turned into a matter of management, and perceived. Nevertheless, the indigenous tribes have never stopped resisting by asserting their cultural wisdom, defining their own abilities with their own words, and demanding their rights as self-determined peoples. Recognizing this interrelationship is the key to converting the disability concept from an individual weakness to a historical force that annihilates Indigenous self-rule. Accordingly, the scope of this chapter is to showcase the Indigenous knowledge, environmental wisdom, and community-based concepts of capability as essentials to the creation of more just, culturally appropriate, and healing approaches to disability and resilience in Oceania.

6.2 Disability and Indigenous Enunciates in Oceania

The demand for recognition of indigenous peoples' knowledge systems is growing, along with the attention these systems receive at global, regional, national, and local levels. Research from the Indigenous Pacific is no different (Sanga, 2004). According to recent research, indigenous groups view the term "disability" as foreign and somewhat at odds with their customs about disabilities (Connell, 2011; Hickey, 2008a, 2008b; King, 2010). According to research done in Western Australia, an Indo-Pacific region, about 20 years ago by Ariotti (1999), the Anangu people accepted humanity's diversity and differences and valued individuality rather than perceiving disabilities. Meekosha (2011) has therefore called on scholars and decision-makers in Australia to pay attention to indigenous peoples' conceptions of disability, as these are not included in the country's current laws and policies.

The Māori community in New Zealand has also stated that disability needs to be conceptualized in a more assenting and cohesive way than is the case in Western scholarship, which is consistent with the findings of research on Anangu people. According to Māori, humanity is a special and interconnected phenomenon with interrelations at its core, connecting the past, present, and future through spiritual well-being and land attachment. Therefore, it is not easy to translate the interdependence of heritages, time, and geography to the Western conception of humankind. Because of this indigenous Māori abstraction, which values individuality and differences, disability is considered a normal aspect of life. Because indigenous inquiry encompasses non-empirical and non-generalizable characteristics, this method creates a significant gap between the indigenous sense of inimitability and the non-indigenous concepts of impairment and disability (Velarde, 2018).

According to Gotto's (2009) investigation of the integration of people with intellectual disabilities into indigenous communities in Mexico, these individuals were accepted and valued members who were acknowledged for their accomplishments rather than their disabilities. Indigenous traditional beliefs strive to accept people as humans first, with a disinclination to define them as divergent, impaired, "deficient," or disabled, which is consistent with research from Mexico and instances from Australia and New Zealand. Shakespeare (1996) noted cautiously that although being disabled is a specific social identity of a minority and having an impairment may be a shared experience, indigenous people do not embrace being impaired as a social identity. Additionally, Watson (2002) contended that this disparity is not exclusive to indigenous peoples and that disabled members of non-indigenous communities have not always embraced the word "disability." Watson (2002) questioned the idea of identity in disability scholarship in his paper "Well, I know this is going to sound very strange to you, but I don't see myself as a disabled person: identity and disability." He contended that all "disabled" people share one crucial characteristic, which is that they are all subject to subjugation heedlessly of how they identify. Shakespeare's claims have been strengthened by Riddell and Watson (2014), who assert that culture can both oppress and liberate people with disabilities. In this way, it is clear that the global north's political strategy for the emancipation of PwD has, in some ways, deepened the gap between the north and the south and strengthened repressive beliefs.

Barker and Murray (2010) provide additional insight into indigenous identity by explaining that for some indigenous people, determining normalcy in wellness and health depends on whether an individual is in steadiness with their social connections, family, spirituality, and ancestral attachment to the land. Responsiveness to cultural needs, which is part of cultural proficiency, is tied to this recognition of indigenous needs. However, rather than developing theories or discourses that are pertinent to disability based on the experiences of indigenous people, addressing the needs of indigenous people has been restricted to examining results (Bickenbach, 1999). The development of a constructivist approach to disability is pertinent given the emphasis on culture in indigenous populations, as it integrates indigenous culture and life experiences in the postcolonial era (Snyder & Mitchell, 2007). By addressing the diversity, distinctiveness, and history of indigenous regions, tribes, and communities worldwide, such an approach makes indigenous identity, self-determination, and culture the catalyst for the manumission of indigenous peoples in the global south (Minde, 2008; Durie, 1998).

Hickey (2008a, 2008b) used the notion of "othering" (Foucault, 1972) to highlight the significance of indigenous identity and to explain how non-indigenous people have viewed disabled indigenous identities negatively. Foucault (1972) first proposed this idea after interacting with vulnerable individuals who were viewed as "other" rather than valuable. Another facet of "othering" and indigenous prejudice is the absence of research on how disability is viewed in indigenous cultures (Creamer, 2009). Lavallee and Poole (2010) have bickered in favour of this viewpoint that knowledge of indigenous peoples' systems and beliefs has been overlooked and that it ought to be added to, expanded upon, and contributed to modern disability inquiry.

Based on the Samoan perspective, one major obstacle to inclusion and acceptance of disability is the lack of societal congruity. It was found that at least three recognized disability frameworks—the moral, medical, and social disability models—are incorporated into Samoan disability conceptualizations. From a Samoan perspective, a spiritual concept of blessings and curses is incorporated into a moral disability model. Medical viewpoints have had a significant impact on the creation of disability language as well as the conceptualization of disability as a sickness. Although the social disability model has been supported as a tool for policy formulation, it has faced opposition from deeply ingrained societal attitudes derived from moral and medical models. According to the research's findings, Samoan culture is still in a state of ideological transformation when it comes to how it views disability. On the basis of the cultural conceptualization of inclusion within the Samoan framework, identity formation integrates ideas of honor, belonging, and familial contribution. Contributing to the group's success is a highly valued quality that strengthens a sense of identification and belonging and increases solidarity, ability, and prestige. According to this study, people with disabilities are largely excluded from required contributions to the extended family. They are actively discouraged from doing so to prevent social shame because of the prevalent Samoan conceptions of disability, which are morally or medically based. Securing cultural identity, status, equality, and inclusion is greatly impacted by this (Picton et al., 2016).

The combined perspectives of the Indigenous people of the Pacific and other regions exhibit that the interpretation of disability is firmly based on cultural views, interdependent identities, and social responsibilities, rather than the biomedical or deficit models that the Global North has imposed. The Anangu, Māori, Mexican Indigenous, and Samoan communities are different from each other in one aspect—they all see disability through connectedness, spirituality, land-based identity, and group contribution. However, colonialism, Western disability models, and processes of "othering" have pushed these culture-based understandings aside, and the labels that have replaced them continue to fragment identity, undermine belonging, and keep people socially excluded. From our perspective, it is necessary to first acknowledge Indigenous epistemologies as valid sources of theory, secondly, to accept the multiplicity of Indigenous experiences, and thirdly, to resist the assumption that Western disability constructs are universal, in order to move towards truly inclusive and decolonized disability scholarship. By placing Indigenous knowledge systems—and the cultural, historical, and ecological wisdom they contain—of disability reinterpreted in ways that restore dignity, affirm identity, and support self-determination for the Indigenous peoples in Oceania and across the global south.

6.3 Disaster Risks and Management Among Pacific Indigenous Disability

According to the Economic and Social Commission for Asia and the Pacific, the Pacific is among the world's most disaster-prone areas. This region's small island nations are particularly vulnerable to earthquakes, tsunamis, volcanic eruptions, storm surges, floods, and cyclones. Furthermore, these nations' small size, scarcity of natural resources, limited economic diversity, and remoteness from major markets all contribute to their financial instability. Many individuals believe that disasters are "extreme natural events that are outside our control, and people affected are victims of a maleficent nature." But according to other research, structural problems in communities lead to an incongruent degree of vulnerability for some groups, including PwD. The degree to which people, communities, and nations experience vulnerability or resilience is often determined by the interconnections among multiple structural challenges. Vulnerable groups, including PwD, experience extra difficulties during and after catastrophes due to these already challenging circumstances. PwD may encounter difficulties during phases of disasters, whether they are intellectual, physical and mobility-related or sensory, or psychosocial. Additionally, underlying social

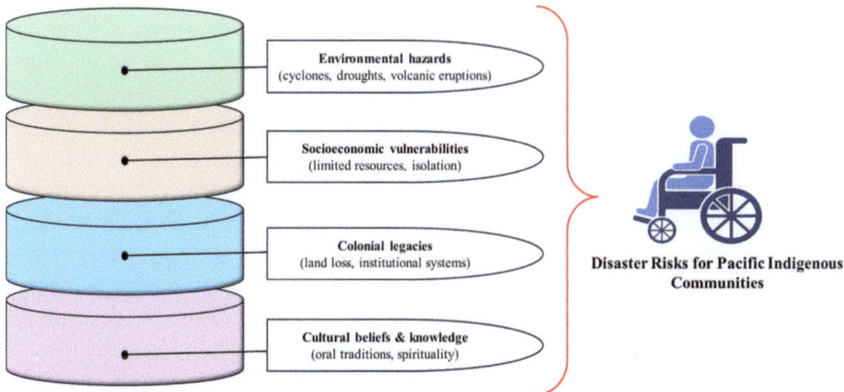

Fig. 6.1 Layers of environmental, social, historical, and cultural factors shaping disability-related vulnerability in disaster contexts. This figure illustrates how multiple interconnected factors contribute to the heightened vulnerability of persons with disabilities during disasters. Environmental hazards such as cyclones, droughts, and volcanic eruptions interact with socioeconomic challenges like limited resources and geographic isolation. These vulnerabilities are further compounded by colonial legacies, including land dispossession and institutional structures that shape present-day inequities. Additionally, cultural beliefs, oral traditions, and spiritual practices influence community responses and perceptions of disability. Together, these layered factors create a complex risk environment that disproportionately affects individuals with disabilities

stigma, discrimination, and isolation can exacerbate both immediate and long-term impacts. Research has largely ignored PwD's experiences of catastrophes, despite—or perhaps because of—their marginalized status and vulnerability in society (King et al., 2019). The various factors affecting the disability vulnerability in the disaster context are depicted in Fig. 6.1.

Hidden under the layers of socio-ecological environments, cultural beliefs, and historical encounters with colonizers and Western authorities, the Indigenous communities scattered throughout the Pacific are facing multi-layered disaster risks. Hazards like cyclones, droughts, floods, volcanic eruptions, and earthquakes are taking a heavy toll on the already subsistence-based livelihoods, thus making the underprivileged groups—including PwD—more exposed to disasters. The multi-faceted disaster menace scenarios of Pacific Indigenous communities are not solely the result of their ecological conditions, but also of two factors: their subsistence modes and cultural viewpoints on environmental hazards. The study of the different communities in the Southwest Pacific, mainly Papua New Guinea and Vanuatu, shows that these communities have to live with the frequent cycles of cyclones, droughts, earthquakes, and volcanic eruptions which are, on the one hand, significant threats to their living and cultural change by the direct route and, on the other, a challenge to the survival of cultural heritage through the indirect route. In the past, these risks have been mitigated through Indigenous peoples' coping systems, e.g., oral storytelling, intergenerational knowledge transfer, spiritual interpretations, and social memory, which have always been community-based risk-reduction methods.

Such systems are particularly significant in the context of PwD because older adults, the less mobile, or those who rely on collective support networks, are more dependent on the community's protection (Jackson, 2021).

Research conducted by the Bedamuni in Papua New Guinea has shown that Indigenous worldviews have a powerful influence on disaster perception and management. The Bedamuni people recognize droughts as a danger but mainly associate them with spiritual and moral factors, which they attribute to cultural narratives and cosmological beliefs. Their historical adaptation to harsh conditions included relocating food production, among other measures, and even moving plants to wet areas, which protects the entire community, including those with mobility issues. These resilience strategies, firmly rooted in the local community, are a testament to how Indigenous systems could assist in a more inclusive disaster preparedness process (Fair, 2019; Jackson, 2020).

A striking mix of traditional and modern approaches to disaster risk reduction can be seen in the Indigenous people's view and management of disasters on Emae Island, Vanuatu. Cyclones, droughts, and climate change are concerns for the community, influenced by a combination of recent disasters, past removals, and government and NGO scientific communication. The impacts of disasters are increasingly linked to climate change, yet the community still relies on traditional management methods, such as lagoon and reef management, considering them eco-friendly and as food reserves after a disaster. These systems are, in fact, very significant for the food security of vulnerable groups, and even more so during disaster recovery. In Vanuatu, disaster management institutions often adopt Western, event-centered risk interpretations, leading to the natural disaster scenario being deemed unavoidable. The institutions' framing puts emergency response above the systemic vulnerabilities that need to be addressed. The neglected needs of such disabled people can be in the form of having their communication systems, evacuation plans, and even decision-makers represent them. Moreover, the narratives that the donor directs have it that Vanuatu is a country that is "disaster-prone" by nature, and, thus, these policies and funding that focus on response instead of resilience are made, and money flows in that direction (Calandra, 2020; Campbell & Barnett, 2010; Fair, 2019; Jackson et al., 2017, 2019; Macdonald, 2019; Tomlinson, 2006).

The islanders of Emae, with a much longer interface with Western institutions, considered calamities or disasters in the same way as the world's leading professionals (i.e., as abnormal, extreme events). On the other hand, although the Bedamuni were quickly changing their ways, they still mostly relied on traditional beliefs to interpret the hazards, and they had historically treated the limited food supply due to hazards and hunger gaps as usual. The society of Emae has extreme customs, and they pray to the ancestors and believe in nature spirits; however, recognizing large-scale disasters as a natural process that is connected to climate change is becoming more common among them, whereas the Bedamuni people widely regard hazards as being caused by spirits and have linked them to their existing eschatological beliefs. Disaster management practitioners in Vanuatu always characterized disasters as being inexorable and anticipatable. The research demonstrated that religion, a distinguishing feature of both societies, does not act as a barrier to their carrying out meaningful risk-reduction activities. The different moral and spiritual causes of hazards across societies must be understood by external DRR stakeholders. The people of Emae and Bedamuni were able to see their livelihood risks from multiple angles, but outside investments will only be successful if local nuances concerning the timing of disasters are respected (Jackson, 2021).

The study conducted by Elisala and his collaborators in 2020 in Tuvalu revealed that capacity-building, communication accessibility, economic conditions, and cultural channels predominantly influenced disaster preparedness in the Pacific Islands. The PwDs were adamant that they should be ready as a group—this mainly involved inclusive evacuation drills, training for evacuation personnel to become acquainted with disabilities, and providing continuous access to disaster information. The researchers noted that such practice is in line with global recommendations, which urge governments and disability organizations to involve PwDs in disaster planning and capacity-building activities (CDC, 2017; Elisala et al., 2020; UNISDR, 2015). The presence of PwDs in neighborhood discussions and the creation of government regulations were pointed out as key aspects to their empowerment, and this is in line with the overall tendency in the literature that has shown that participation in decision-making fosters preparedness and facilitates an equitable response to humanitarian aid (Elisala et al., 2020; Lord et al., 2016; WHO, 2013). Another significant problem is efficient communication, which requires applying specific tactics, e.g., working with caregivers, using local language in public notices, providing sign language interpreters, and setting alarms at appropriate distances. In addition to communication problems faced by disabled persons and their need for solutions, research has identified suggested strategies.

The investigation found that the major hindrances to the preparedness of PwDs were, namely, unemployment, absence of planning on the part of the family and government, social and attitudinal discrimination, and physical and communication barriers that hampered timely evacuation (CDC, 2017; Elisala et al., 2020; FEMA, 2014; WHO, 2013). It was noted that unemployment affected over 90% of the respondents, consistent with the low socioeconomic status among PwDs reported, which is prevalent both regionally and globally, and with the difficulty in buying necessary disaster supplies. Without sufficient financial and social support, PwDs are the ones who suffer most during disasters, especially when they need to be evacuated. A lot of the respondents admitted that they would require help to get out, which is in line with earlier studies that pointed out the strong dependence of disabled people on caregivers or community members during emergencies. Such dependence could turn out to be a disadvantage in itself when help is not available or arrives late, as some related studies have noted. These vulnerabilities are further burdened by the oppression of disabled children and youth in the Pacific communities, which makes their participation and visibility in preparedness planning very limited (Powell & Gilbert, 2006; Tavola, 2012; WHO, 2013).

Even with these restrictions, the research still shows that the motivation behind the disabled community's disaster preparations was mainly determined by cultural and spiritual factors, such as family safety, traditional warning systems (Te Valo), praying, and individual experience with previous disasters. Trusted information sources, such as government agencies, meteorological offices, and local alert systems (Becker et al., 2017; Steelman & McCaffrey, 2013), play a significant and consistently recognized role in preparedness within the disaster research field. Participants' understanding of types of disasters, coping methods, and both traditional and scientific interpretations was a powerful factor in preparedness, aligning with the broader literature, which recognizes knowledge as a crucial attribute of resilience. Past disasters experienced by individuals with disabilities also played a role in shaping their perceptions about risk, vulnerability, and preparedness tactics, harboring clues that the experience of disaster often determines the response, no matter the disability type. The results of the study call for Pacific disaster management practices to be grounded in the day-to-day experiences, knowledge, and decision-making input of people with disabilities, thereby ensuring community resilience through the inclusion of people with disabilities in preparedness activities (Elisala et al., 2020).

6.4 Disability Inclusive Knowledge: Challenges and Possible Threats

The Pacific Islands are the most affected by global climate change, and the extinction of Indigenous disabled-inclusive knowledge and practices is the most direct threat to the land. The destruction of the ecosystems and extinction of the flora and fauna that are connected to the culture of the people is going to be the main reason for the loss of the heritage of the island communities in particular Kiribati, Tuvalu, Marshall Islands, and Fiji which are right on the crossroads where the climate change scenarios are very evident in the form of sea-level rise, more violent storms, and faster coastal erosion. This means the islanders are literally losing ground—good soil, clean water, and the places where they used to live—thus, movement and resettlement are inevitable. To illustrate, the people of Kiribati are already suffering as the sea has started to mix with their freshwater, and coastal erosion is destroying their agricultural land. The case of Fiji shows that when local, traditional methods reliant on Indigenous knowledge are not incorporated into the implementation of climate change adaptation measures, one is often less successful. Not only does the confiscation of traditional lands literally drive the Indigenous people from their places, but concurrently, it makes their culture and social life unfeasible. For people with disabilities, a traditional land is not just a resource; it is a place where all the care-giving networks and practices from the past live together and continue to evolve. With environmental changes pushing families to migrate to either scattered urban areas or different islands, the very basis of support among relatives is diminishing, which in turn affects the disabled persons depending on community-based care. It is therefore necessary to act quickly to integrate the growing awareness of climate change with the uncovering of indigenous knowledge while disability-inclusive aspects are fully preserved (Cory, 2024; Haque, 2025; Johnson et al., 2021; Nunn & Kumar, 2019; Park et al., 2024; Yamamoto, 2020).

The oral traditions, rituals, and cultural practices of the Indigenous people in the Pacific are primarily the means by which their knowledge systems are passed down, making them susceptible to the loss of language and the decline of knowledge keepers. The history of colonization, forced assimilation, and the prioritization of globalized education have harmed the Pacific languages and have led to their extinction. This loss not only silences their wisdom spanning centuries but also the Indigenous languages that often provide means to express differences and abilities beyond Western individualistic notions. For instance, in the Solomon Islands, women and older people play an important role in sharing knowledge about and practices that are advantageous for people with disabilities, with mangrove foodscapes providing necessary resources for people with mobility disabilities. When the older generation dies without passing on their knowledge, or when the younger generation lacks the means or interest to learn their ancestors' languages, traditional healing practices, and inclusive educational systems suffer. A qualitative study from Fiji cites the dangers of overlooking traditional and local coping mechanisms in climate change adaptation strategies, thereby making the community's capacity to address disability inclusivity more dependent on the presence of knowledge keepers. This situation gives birth to gaps in the understanding of Indigenous methods that extend across generations, making the communities weaker in times of crises and less capable of supporting individuals with disabilities (Bruckner & Paia, 2025; Cory, 2024; Nunn et al., 2024).

With regards to Colonial Legacies, Structural Marginalization, and Exclusion, the colonial process forced Western law, education, and religion upon Pacific societies, generally disregarding Indigenous knowledge and further excluding people with disabilities from society. Colonialism went on to monopolize the structure of care in Fiji, Samoa, Vanuatu, and Papua New Guinea by replacing the community care model with an institutional and medical one, which only aggravated the isolation of the disabled population. Developmental policies that were imported seldom consider Indigenous people's views on kinship and interdependence and tend to be focused on "rights" frameworks that sometimes diverge from local practices. The newly set-up colonial regimes rule with an iron hand, and their marginalization extends through educational programs, policy-making, and resource allocation. For example, disaster mitigation in Papua is said to be more effective when traditional leaders are involved, but such leaders are often left out due to external intervention and policy priorities. Indigenous rulers and disabled individuals are often not able to connect with the political power areas directly, which contributes to their exclusion from the governance of their lives. This kind of challenge is inviting the decolonization of knowledge production and governance in the Pacific to be the only way out, accepting the Indigenous peoples as the guardians of disability-inclusive systems and thus making their voices heard in climate change adaptation and development planning (Cory, 2024; David et al., 2021; Fitriani et al., 2024; Haque, 2025; Nunn & Kumar, 2019; Nunn et al., 2024).

6.5 Disability Inclusive Resilience: The Opportunities

Acknowledging Indigenous people's ways of knowing is one of the most effective means of creating a stronger and resilient population of disabled people in the Pacific communities. The whole community, caring for and including all members, including persons with disabilities, has always been the very core of such practices through extended family, communal land use, and local innovative technologies. For example, the resource management practices encompassing agroforestry and sustainable harvesting have been sanctioned over the generations, thus offering a wealth of experience in climate adaptation and livelihoods that are inclusive of people with disabilities. Knowledge transfer from the old to the young generation occurs through storytelling, apprenticeships, and ceremonies, which, in turn, promote cultural continuity and, at the same time, legitimize the position of the elderly and Indigenous people with disabilities as carriers of knowledge. Community-driven adaptation initiatives in areas like Vanuatu have demonstrated that when Indigenous governance and traditional practices are prioritized, adaptation is not only more efficient but also more inclusive (Nunn et al., 2024; Westoby et al., 2020).

In the Pacific, the most powerful resilience-building activities are those that originate with local people, not those imposed from outside. Community-led disaster risk reduction (DRR) projects have been successful on many islands by incorporating traditional knowledge and the participation of local leaders, people with disabilities, and extended kin networks. In Vanuatu, community-led adaptation planning has shifted the term from community-based to locally led, thereby enhancing the voices and necessities of PwD and ensuring that solutions are contextually appropriate and accepted by the whole village. Collaborations between external organizations and local communities that respect cultural protocols have been very fruitful. The partnerships are productive because they enable knowledge exchange, co-design, and mutual accountability, thereby improving disaster readiness and increasing accessibility for PwD (Ali et al., 2021; Westoby et al., 2020).

The process of making Indigenous populations more resilient will not be possible unless the governments, researchers, and NGOs cooperate with Indigenous communities, and the local definitions of inclusion and disability are accepted. Local networks such as the Moanaroa Pacific Research Network in Aotearoa/New Zealand illustrate how Indigenous views and authority can shape research and policy-making in ways that mirror the social, cultural, and ecological aspects of Pacific peoples. Integrated care models, particularly those developed in collaboration with Pacific families, address health, education, and social inclusion simultaneously, which is why they support not just individuals but whole family units. They provide a different approach to disability services and policies, combining Western and Indigenous principles to achieve a more inclusive outcome (Cammock et al., 2022; Enari et al., 2024; Matapo & Enari, 2024; Palinkas et al., 2022).

The smaller-than-expected reservation is one of the best chances to give a voice to Indigenous leaders, persons with disabilities, and their allies by involving them in research, policy-making, and education from the very beginning. Training, scholarships, and participatory research are among the capacity-building activities that ensure the influence of Indigenous peoples' voices in decision-making. The educational activities that reveal the knowledge of both Indigenous and Western viewpoints are the ones that help the revival of Indigenous languages, histories, and methods of dealing with disability, which are lessening stigma and encouraging the community to speak for itself. Collaborations in the regional climate health and research-practice networks are also working on health equity, mental health resilience, and environmental adaptation (Matapo & Enari, 2024; Palinkas et al., 2022).

6.6 Conclusion

The connection between Indigenous peoples, disability, and ecological wisdom in Oceania is made clear in this chapter, which also points out that the disability of the Indigenous people cannot be seen in isolation from the larger picture of historical and political factors of colonial removal, cultural suppression, and Western standardization of normalcy. Communities in the Pacific area continue to fight to get rid of disability models that focus on deficits, presenting a concept of ability grounded in relationships with land and community, thus honoring one's cultural identity, spiritual wholeness, and interdependence. These Indigenous ways of knowing show that disability is not only an individual affliction but rather a condition influenced by social, environmental, and historical factors, including the legacies of forced assimilation, educational exclusion, and structural marginalization. On one hand, the overlap of disability and disaster risk clearly points out the necessity of disaster management that is culturally grounded and inclusive of people with disabilities. As climate change advances, the danger of destruction in the environment, people with disabilities among the Indigenous population are suffering the most because of the very same reasons, among others, like unequal distribution of wealth, loss of land due to colonization, communication obstacles, and policy marginalization. However, the communities of the Pacific include rich ecological wisdom such as traditional resource management, oral knowledge transmission, spiritual interpretations of environmental change, and community-based care, which can be very useful for disaster risk reduction and thereby strengthen resilience. The full partaking of PwD in community governance, early warning systems, evacuation planning, and human resource training is vital for heightening the community's preparedness and achieving fairness in outcomes.

References

Ali, T., Buergelt, P. T., Paton, D., Smith, J. A., Maypilama, E. L., Yuŋgirrŋa, D., & Gundjarranbuy, R. (2021). Facilitating sustainable disaster risk reduction in Indigenous communities: Reviving

Indigenous worldviews, knowledge and practices through two-way partnering. *International Journal of Environmental Research and Public Health, 18*(3), 855. https://doi.org/10.3390/ijerph18030855

Annamma, S. A., Ferri, B. A., & Connor, D. J. (2018). Disability critical race theory: Exploring the intersectional lineage, emergence, and potential futures of DisCrit in education. *Review of Research in Education, 42*(1), 46–71. https://doi.org/10.3102/0091732X18759041

Ariotti, L. (1999). Social construction of an Angu disability. *Australian Journal of Rural Health, 7*(4), 216–222. https://doi.org/10.1111/j.1440-1584.1999.tb00460.x

Barker, C., & Murray, S. (2010). Disabling postcolonialism: Global disability cultures and democratic criticism. *Journal of Literary & Cultural Disability Studies, 4*(3), 219–236. https://doi.org/10.3828/jlcds.2010.20

Becker, J. S., Paton, D., Johnston, D. M., Ronan, K. R., & McClure, J. (2017). The role of prior experience in informing and motivating earthquake preparedness. *International Journal of Disaster Risk Reduction, 22*, 179–193. https://doi.org/10.1016/j.ijdrr.2017.03.006

Bickenbach, J. E. (1999). Minority rights or universal participation: the politics of disablement. In *Disability, divers-ability and legal change* (pp. 101–115). Brill Nijhoff.

Bruckner, H. K., & Paia, M. T. (2025). From mangroves to womangroves to feminist foodscapes:(en) gendering research on indigenous food livelihoods in the Solomon Islands. *Agriculture and Human Values, 42*(1), 507–525. https://doi.org/10.1007/s10460-024-10634-8

Calandra, M. (2020). Disasta: Rethinking the notion of disaster in the wake of Cyclone Pam. In *Anthropological forum* (Vol. 30, No. 1–2, pp. 42–54). Routledge. https://doi.org/10.1080/00664677.2019.1647826

Cammock, R., Boon-Nanai, J., Uasike Allen, J. M., Keung, S., Enari, D., Vaka, S., Ahio, L., Cammock, R., Nanai, J. M. B., Uasike Allen, J. M., Te, Aronui, W., Rau, T. M., & Tautolo, E. S. (2022). Promoting Pacific Indigenous research perspectives and pedagogy within postgraduate health research course development. *The Journal of the Polynesian Society, 131*(4), 427–451. https://doi.org/10.15286/jps.131.4.427-452

Campbell, J., & Barnett, J. (2010). Climate change and small island states: Power, knowledge and the South Pacific. *Routledge*. https://doi.org/10.4324/9781849774895

CDC. (2017). *Emergency preparedness: Including people with disabilities*. Centre for Diseases Control and Prevention; 2017. http://www.cdc.gov/ncbdd/disabilityandhealth/emergencypreparedness.html

Connell, R. (2011). Southern bodies and disability: Re-thinking concepts. *Third World Quarterly, 32*(8), 1369–1381. https://doi.org/10.1080/01436597.2011.614799

Cory, J. (2024). 'We are of one ecology': How indigenous Pacific Islander poetics map anthropogenic climate change. *Comparative American Studies an International Journal, 21*(3–4), 217–230. https://doi.org/10.1080/14775700.2025.2451451

Creamer, D. B. (2009). Disability and Christian theology: Embodied limits and constructive possibilities. *OUP USA*. https://doi.org/10.1093/acprof:oso/9780195369151.001.0001

David, C. G., Hennig, A., Ratter, B. M., Roeber, V., Zahid, & Schlurmann, T. (2021). Considering socio-political framings when analyzing coastal climate change effects can prevent maldevelopment on small islands. *Nature Communications, 12*(1), 5882. https://doi.org/10.1038/s41467-021-26082-5

Durie, M. H. (1998). *Te Mana, Te Kāwanatanga: The politics of self determination*. Oxford University Press.

Elisala, N., Turagabeci, A., Mohammadnezhad, M., & Mangum, T. (2020). Exploring persons with disabilities preparedness, perceptions and experiences of disasters in Tuvalu. *PLoS ONE, 15*(10), Article e0241180. https://doi.org/10.1371/journal.pone.0241180

Enari, D., Matapo, J., Ualesi, Y., Cammock, R., Port, H., Boon, J., Refiti, A., Fainga'a-Manu, I., Fainga'a-Manu Sione, I., Thomsen, P., & Faleolo, R. (2024). Indigenising research: Moanaroa a philosophy for practice. *Educational Philosophy and Theory, 56*(11), 1044–1053. https://doi.org/10.1080/00131857.2024.2323565

Fair, H. (2019). From apathy to agency: Exploring religious responses to climate change in the Pacific Island region. *Dealing with climate change on small islands: Towards effective and sustainable adaptation?*, 175. https://doi.org/10.17875/gup2019-1216

FEMA. (2014). *Preparedness in America: research insight to increase individual, organization and community action.* Federal Emergency Management Agency. http://www.fema.gov/media-librarydata/f9728f1bf52a691b2602d7d49cd423a9/20130904_Preparedness+in+America_FINAL_508.pdf

Fitriani, F., Tjilen, P. A., Betaubun, M., Suhendra, A. F. (2024). Regional head leadership model in mitigating climate change in Papua. *Journal of Infrastructure, Policy and Development, 8*(15), 9114. https://doi.org/10.24294/jipd9114

Foucault, M. (1972). *The archaeology of knowledge: Translated from the french by AM Sheridan Smith.* Pantheon Books.

Gilroy, J., Uttjek, M., Lovern, L., & Ward, J. (2021). Indigenous people with disability: Intersectionality of identity from the experience of Indigenous people in Australia, Sweden, Canada, and USA. *Disability and the Global South, 8*(2), 2071–2093.

Goodley, D. (2017). Dis/entangling critical disability studies. *Culture–theory–disability*, 81. https://doi.org/10.14361/9783839425336-008

Goodley, D. (2024). *Disability studies: An interdisciplinary introduction.* https://doi.org/10.1111/j.1755-618X.2011.01272.x

Gotto, G. (2009). Persons and nonpersons: Intellectual disability, personhood, and social capital among the Mixe of Southern Mexico. *Disabilities: Insights from across fields and around the world* (Vol. 1, pp. 193–209).

Haque, A. (2025). *Resilience on the frontlines: Navigating climate change in the Pacific Islands.* Available at SSRN 5382395. https://doi.org/10.64142/jebs.1.1.9

Hickey, H. (2008). *Claiming spaces: Maori (indigenous persons) making the invalid valid.* https://www.academia.edu/68366573/Claiming_spaces_Maori_indigenous_persons_making_the_invalid_valid

Hickey, S. J. (2008). *The unmet legal, social and cultural needs of Māori with disabilities* (Doctoral dissertation). The University of Waikato. https://hdl.handle.net/10289/2571

Ineese-Nash, N. (2020). Disability as a colonial construct: The missing discourse of culture in conceptualizations of disabled Indigenous children. *Canadian Journal of Disability Studies, 9*(3), 28–51. https://doi.org/10.15353/cjds.v9i3.645

Jackson, G. (2020). The influence of emergency food aid on the causal disaster vulnerability of Indigenous food systems. *Agriculture and Human Values, 37*(3), 761–777. https://doi.org/10.14264/c3d6353

Jackson, G. (2021). Perceptions of disaster temporalities in two Indigenous societies from the Southwest Pacific. *International Journal of Disaster Risk Reduction, 57*, Article 102221. https://doi.org/10.1016/j.ijdrr.2021.102221

Jackson, G., McNamara, K. E., & Witt, B. (2019). *Conducive and hindering factors for effective disaster risk reduction in Emae Island, Vanuatu.* Contributing Paper to GAR 2019. https://www.preventionweb.net/publications/view/66187

Jackson, G., McNamara, K., & Witt, B. (2017). A framework for disaster vulnerability in a small island in the Southwest Pacific: A case study of Emae Island, Vanuatu. *International Journal of Disaster Risk Science, 8*(4), 358–373. https://doi.org/10.1007/s13753-017-0145-6

Johnson, F., Higgins, P., & Stephens, C. (2021). Climate change and hydrological risk in the Pacific: A Humanitarian Engineering perspective. *Journal of Water and Climate Change, 12*(3), 647–678. https://doi.org/10.2166/wcc.2021.277

Joona, T. (2012). *ILO convention no. 169 in a Nordic context with comparative analysis: An interdisciplinary approach.* fi= Lapin yliopistokustannus| en= Lapland University Press. https://research.ulapland.fi/fi/publications/ilo-convention-no-169-in-a-nordic-context-with-comparative-analys/

Ka Toni, M. (2007). *The production of an appropriate and culturally sound isiXhosa translation of the International Classification of Functioning, Disability and Health (ICF) Checklist*. http://hdl.handle.net/11427/11635

King, J. A. (2010). *Weaving yarns: the lived experience of Indigenous Australians with adult-onset disability in Brisbane* (Doctoral dissertation). Queensland University of Technology. https://doi.org/10.1080/09687599.2013.864257

King, J., Edwards, N., Watling, H., & Hair, S. (2019). Barriers to disability-inclusive disaster management in the Solomon Islands: Perspectives of people with disability. *International Journal of Disaster Risk Reduction, 34*, 459–466. https://doi.org/10.1016/j.ijdrr.2018.12.017

Lavallee, L. F., & Poole, J. M. (2010). Beyond recovery: Colonization, health and healing for Indigenous people in Canada. *International Journal of Mental Health and Addiction, 8*(2), 271–281. https://doi.org/10.1007/s11469-009-9239-8

Lord, A., Sijapati, B., Baniya, J., Chand, O., & Ghale, T. (2016). *Disaster, disability, & difference*. https://doi.org/10.13140/RG.2.2.30295.93609

Lovern, L., & Locust, C. (2013). Native American communities on health and disability: A borderland dialogues. *Springer*. https://doi.org/10.1057/9781137312020

Macdonald, F. (2019). 'God was here first': Value, hierarchy, and conversion in a Melanesian Christianity. *Ethnos, 84*(3), 525–541. https://doi.org/10.1080/00141844.2018.1456477

Matapo, J., & Enari, D. (2024). Welcome from Jacoba Matapo and Dr Dion Enari, Chair Associate Professor and Co-Chair of Moanaroa Pacific Research Network. *Rangahau Aranga: AUT Graduate Review, 3*(1). https://doi.org/10.24135/rangahau-aranga.v3i1.225

Meekosha, H. (2011). Decolonising disability: Thinking and acting globally. *Disability & Society, 26*(6), 667–682. https://doi.org/10.1080/09687599.2011.602860

Minde, H. (Ed.). (2008). *Indigenous peoples: Self-determination, knowledge, indigeneity*. Eburon Uitgeverij BV.

Nunn, P. D., Kumar, R., Barrowman, H. M., Chambers, L., Fifita, L., Gegeo, D., Gomese, C., McGree, S., Rarai, A., Cheer, K., Esau, D., 'Ofa Fa'anunu, Fong, T., Fong-Lomavatu, M., Geraghty, P., Heorake, T., Kekeubata, E., … & Waiwai, M. (2024). Traditional knowledge for climate resilience in the Pacific Islands. *Wiley Interdisciplinary Reviews: Climate Change, 15*(4), Article e882. https://doi.org/10.1002/wcc.882

Nunn, P., & Kumar, R. (2019). Measuring peripherality as a proxy for autonomous community coping capacity: A case study from Bua Province, Fiji Islands, for improving climate change adaptation. *Social Sciences, 8*(8), 225. https://doi.org/10.3390/socsci8080225

Oliver, M. (2017). Defining impairment and disability: Issues at stake. In *Disability and equality law* (pp. 3–18). Routledge. https://www.um.es/discatif/PROYECTO_DISCATIF/Textos_discapacidad/00_Oliver2.pdf

Palinkas, L. A., O'Donnell, M., Kemp, S., Tiatia, J., Duque, Y., Spencer, M., Basu, R., Del Rosario, K. I., Diemer, K., Doma Jr, B., Forbes, D., Gibson, K., Graff-Zivin, J., Harris, B. M., Hawley, N., Johnston, J., Lauraya, F., Maniquiz, N. E. F., Marlowe, J., McCord, G. C., Nicholls, I., Rao, S., Kim Saunders, A., … & Wong, M. (2022). Regional research-practice-policy partnerships in response to climate-related disparities: Promoting health equity in the Pacific. *International Journal of Environmental Research and Public Health, 19*(15), 9758. https://doi.org/10.3390/ijerph19159758

Park, Y. S., Kim, H., & Kim, T. (2024). Changing vulnerability to climate change and health impacts in an Outer Island in Kiribati: A qualitative study. *국제개발협력연구, 16*(4), 115–127. https://doi.org/10.32580/idcr.2024.16.4.115

Picton, C., Horsley, M., & Knight, B. A. (2016). *Exploring conceptualisations of disability: A talanoa approach to understanding cultural frameworks of disability in Samoa*. https://doi.org/10.5463/dcid.v1i1.502

Powell, R., & Gilbert, S. (2006). *The impact of hurricanes Katrina and Rita on people with disabilities: A look back and remaining challenges*. National Council on Disability.

Riddell, S., & Watson, N. (2014). Disability, culture and identity: Introduction. In *Disability, culture and identity* (pp. 1–18). Routledge. https://doi.org/10.4324/9781315847634

Sanga, K. (2004). *Making philosophical sense of indigenous Pacific research.* https://doi.org/10.26686/wgtn.12839150

Sarivaara, E., Maatta, K., & Uusiautti, S. (2013). Who is indigenous? Definitions of indigeneity. *European Scientific Journal.* https://research.ulapland.fi/en/publications/who-is-indigenous-definitions-of-indigeneity/

Senior, K. (2000). Testing the ICIDH-2 with Indigenous Australians: Results of Field Work in Two Aboriginal Communities in the Northern Territory: a Final Report Prepared for the Australian Institute of Health and Welfare, ICIDH Collaborating Centre and the Department of Health and Family Services. Australian Collaborating Centre.

Shakespeare, T. (1996). Disability, identity and difference. *Exploring the divide*, 94–113. https://www.researchgate.net/publication/237440415_Disability_identity_and_difference

Snyder, S. L., & Mitchell, D. T. (2007). *Cultural locations of disability.* https://doi.org/10.1080/15017410701345365

Steelman, T. A., & McCaffrey, S. (2013). Best practices in risk and crisis communication: Implications for natural hazards management. *Natural Hazards, 65*(1), 683–705. https://doi.org/10.1007/s11069-012-0386-z

Tavola, H. (2012). Addressing inequalities: disability in Pacific island countries. Paper presented online at the 'Addressing inequalities' Global Thematic Consultation.

Tomlinson, M. (2006). Retheorizing mana: Bible translation and discourse of loss in Fiji. *Oceania, 76*(2), 173–185. https://doi.org/10.2307/40332021

UNISDR. (2015). *Sendai framework for disaster risk reduction 2015–2030.* UNISDR. http://www.wcdrr.org/uploads/Sendai_Framework_for_Disaster_Risk_Reduction_2015_2030.pdf

United Nations General Assembly. (2007). *United Nations Convention on the Rights of Persons with Disabilities (UNCRPD).* https://www.un.org/development/desa/disabilities/convention-on-the-rights-of-persons-with-disabilities/convention-on-the-rights-of-persons-with-disabilities-2.htmlUnitedNationsInter-AgencySupportGroup

Velarde, M. R. (2018). Indigenous perspectives of disability. *Disability Studies Quarterly, 38*(4). https://doi.org/10.18061/dsq.v38i4.6114

Watson, N. (2002). Well, I know this is going to sound very strange to you, but I don't see myself as a disabled person: Identity and disability. *Disability & Society, 17*(5), 509–527. https://doi.org/10.1080/09687590220148496

Westoby, R., McNamara, K. E., Kumar, R., & Nunn, P. D. (2020). From community-based to locally led adaptation: Evidence from Vanuatu. *Ambio, 49*(9), 1466–1473. https://doi.org/10.1007/s13280-019-01294-8

World Health Organization. (2002). Towards a common language for functioning, disability, and health: ICF. *The international classification of functioning, disability and health.* https://cdn.who.int/media/docs/defaultsource/classification/icf/icfbeginnersguide.pdf

World Health Organization. (2013). *Guidance note on disability and emergency risk management for health.* WHO. https://www.who.int/publications/i/item/guidance-note-on-disability-and-emergency-risk-management-for-health

Yamamoto, L. (2020). Climate relocation and indigenous culture preservation in the Pacific Islands. *The Georgetown Journal of International Affairs, 21*, 150. https://doi.org/10.1353/gia.2020.0003

Chapter 7
Crisis Disaster Preparedness: Lessons from Local Initiatives

Abstract The report from the Pacific region regarding disaster readiness for people with disabilities focuses on the local incidents in the Pacific region and, at the same time, the major cultural transformations due to the international policy changes, like the Beijing Declaration and Action Plan, the Incheon Strategy, and the Asian and Pacific Decades of People with Disabilities. The ascendance of community-based projects, the involvement of concerned parties, and the alignment with regional frameworks have underlined the importance of traditional knowledge, accessible language, and participatory planning in creating DRR systems that are both more resistant and more inclusive. Notwithstanding the remarkable advancements made in this regard, persons with disabilities are still experiencing exclusion in society brought about by the lack of accessibility, poor data, structural discrimination, and policy gaps. The chapter suggests that the waged efforts, such as whole-of-government and whole-of-society approaches, co-designed solutions, the empowerment of OPD, and the reform of laws and policies in line with the CRPD and the Sendai Framework, should be continued. In the end, the disability-inclusive DRR in the Pacific will call for coordinated action, sustained financing, and the use of culture-based strategies that prioritize persons with disabilities in disaster governance and community resilience.

Keywords Disability · Disaster · Incheon strategy · Risk reduction · Traditional knowledge

7.1 Introduction

Vulnerabilities are circumstances caused by social, physical, environmental, and economic elements or processes that increase a person's, a community's, a property's, or a system's susceptibility to the effects of hazards. Based on paradigmatic variations in how disasters and their management have been conceptualized, the historical evolution of disaster studies can be reduced to a dichotomy (Gaillard & Mercer, 2013). The danger-vulnerability idea serves as the foundation for the dichotomy.

S. M. Thomas and R. Veerabathiran, *Disability, Disaster, and Resilience in Oceania*, SpringerBriefs in Modern Perspectives on Disability Research,
https://doi.org/10.1007/978-981-95-8535-9_7

The "vulnerability paradigm" is based on a human/political ecological explanation, whereas the "hazard paradigm" concentrates on behavioral interpretations (Hewitt, 2019; Kates, 1971; O'Keefe et al., 1976; Wisner & Wisner, 2004). However, it is understood that reducing catastrophe studies to this dichotomy risks oversimplifying a complex body of literature and ideas.

The behavioral geography movement had a significant impact on the hazard paradigm, which is regarded as the first modern paradigm (Gaillard & Mercer, 2013). Different natural hazards were thought to be responsible for the disasters, and the impacted population was either caught off guard or did not know the risks they were subjected to, thus preventing them from the process of adaptation (Kates, 1971). According to Gaillard and Mercer (2013), the peril paradigm, which Hewitt (2019) referred to as a physicalist paradigm, was focused on the physical hazard and mainly provided scientific and technological mitigation strategies like early warning systems, engineered defenses, and other structural methods like land-use planning and building codes. The technocratic approach, which was founded on centralized contemporary nations, showed little regard for the developing world as the hazards paradigm expanded from the United Nations to the developed world (Quarantelli, 1987).

Many people and organizations around the world still consider disasters to be natural rather than the result of socially induced vulnerability, according to Lavell and Maskrey (2014), demonstrating the continued dominance of the hazard paradigm at both national and international levels. The conceptual distinction between the vulnerability and hazard exemplars influences the final implementation of contemporary disaster risk reduction (DRR). Instead of addressing the underlying socioeconomic causes of vulnerability, framing a tragedy as "natural" will result in reactive management. Despite the widespread use of the phrase "natural disasters," numerous authors have emphasized that disasters are a natural part of human society and only occur when threats combine with the social and physical vulnerabilities of an exposed population (Jackson et al., 2017).

The South Pacific is deemed one of the world's most remote and hazardous places. High levels of risk exposure coupled with several vulnerabilities, including low economic growth, a lack of institutional governance and social protection, and a reliance on vulnerable assets and entire industries that are vital to livelihoods, like agriculture and fisheries. On the other hand, climate change goes on to be a substantial risk factor for hazards in the South Pacific as it leads to a whole range of different impacts like rising sea levels, a raise in El Niño-Southern Oscillation (ENSO) cycles frequency, changes in rainfall patterns, hotter weather and more intense cyclones (although there might be a decline in the total number of cyclones) (Jackson et al., 2017). The information gathered from more than 20 years of evaluations and other studies that have employed a variety of techniques and tools has been used to determine the potential effects of climate change on Pacific Island countries (PICs), the resulting vulnerabilities, resilience, and adaptive capacity of natural and human systems to climate change, as well as potential and prioritized adaptation interventions. The particularity of islands, their marine environment, and the inhabitants have driven the development and application of state-of-the-art techniques and tools, and innovative approaches to use the outcomes to direct planning, resource mobilization, policy-making, and practical actions. Despite their efforts, PICs have not yet seen noticeable improvements in the risks and vulnerabilities associated with climate change. There is a need for a larger, more intelligent effort to address climate change at all levels, as climate risks are rising globally and in the Pacific regions (Hay & Mimura, 2013).

Disaster preparedness in the Pacific area calls for a holistic and organized plan that is in harmony with global standards for vulnerability and resilience, especially for those who have the most difficulties, such as those with disabilities. The United Nations' Sendai Framework for Disaster Risk Reduction 2015–2030 has mentioned four key areas for action: (1) knowing disaster risk; (2) enhancing disaster risk governance to deal with disaster risk; (3) financing disaster risk reduction for resilience; and (4) advancing preparedness for active response and for "Build Back Better" in recovery, rehabilitation, and reconstruction. Financial protection is yet another factor pinpointed by the Global Fund for DRR. The Pacific nations must take steps towards all five disaster risk reduction priorities (Edmonds & Noy, 2018).

With that context, enhancing the Pacific's crisis and disaster preparedness would mean specifically moving towards the adoption of risk management methods that are socially grounded, inclusively involve the most vulnerable people, and are locally emergent. The Pacific Island nations, facing an increasing reality of climate-related hazards and underlying economic and social weaknesses, are learning and benefiting greatly from community initiatives and traditional resilience practices in disaster mitigation. The majority of persons with disabilities, who are even more exposed, face considerable communication barriers and are often excluded from the decision-making process, would find this change from non-participatory to inclusive community-based strategies fundamental. Hence, this chapter examines the role of local projects in Samoa, Fiji, Vanuatu, and the Solomon Islands in advancing more inclusive and sustainable preparedness actions. By reviewing these practices, the chapter highlights the competencies that might strengthen the national DRR systems while making sure that disaster readiness is just, culturally suitable, and proficient in tackling the complex hazards that Pacific communities are encountering these days.

7.2 Disability Engagement in Community-Based Disaster Planning

Hemingway and Priestley (2006) point out that, among other things, institutional discrimination, societal systems, and environmental barriers are the main factors that can cause disabled people to be more vulnerable than others to human disasters. This is due to the conventional perception of PwD as "victims" and members of "vulnerable groups" who are defenseless or reliant on others. Some studies also claim that because individuals with disabilities are perceived as "the least worth saving," their needs are neglected in local emergency preparation (Abbott & Porter,

2013). According to the UNISDR (2013) survey, 85.57% of PwD had not participated in community disaster management in their local communities before the crisis. According to this survey and previous research, several potential explanations for their absence exist. Due to social exclusion or isolation, as well as impairment-related factors including hearing or vision impairment, PwD may find it difficult to obtain early warning and lifesaving information (Kailes & Enders, 2007; Robinson & Kani, 2014; UNISDR, 2013). PwD are more vulnerable in natural disaster situations due to structural barriers created by general demographic and social factors like poverty and gender.

PwD must be included in the development of disaster resilience, according to the Sendai Framework for Disaster Risk Reduction 2015–2030. It embodies the concept of an all-of-society approach. It is the first worldwide disaster risk reduction framework to explicitly mention disability inclusion, following its predecessors, the Hyogo Framework for Action and the Yokohama Strategy, which had previously neglected to do so. Article 11 of the UNCRPD, which addresses PwD's rights to protection in emergencies and other risky situations, is in line with the Sendai Framework. According to Stough and Kang (2015), these tools signify a paradigm shift in disaster preparedness from viewing PwD as "vulnerable" to viewing them as "active contributors." PwD can "engage in the assessment of their own vulnerabilities, needs, and capacities to face those needs" when they are empowered to respond to their vulnerability, according to Wisner et al. (2012).

PwD have been the ones who disaster events affected the most. Their chances of death in a disaster are two to four times higher, they face a greater risk of injuries and losing properties, are more difficult to evacuate, and require more medical and social services during the disaster and after. The series of factors that contribute to the increased vulnerability of PwD to disasters is a significant reason PwD remain marginalized in emergency management practice and policy-making. The United Nations Office for Disaster Risk Reduction indicated this as a global challenge in 2014. The activities and policy advocacy related to Global Disability-Inclusive Disaster Risk Reduction (DIDRR) have been instrumental in getting the principles of accessibility, inclusion, and universal design recognized in the Sendai Framework for Disaster Risk Reduction (SFDRR) 2015–2030. DIDRR involves the collective responsibility of diverse stakeholders who are cooperating to recognize and eliminate risks that prevent PwD from being safe before, during, and after disasters. However, the local authorities and rescue professionals are still confronted with the problem of sharing responsibility among the local government, emergency personnel, PwD, and the services that support them. The sharing of responsibility should not only be discussed but also translated into action through methods, tools, and programmatic guidance that put the PwD and their support needs at the core of emergency management (Villeneuve, 2022).

The Person-Centered Emergency Preparedness (P-CEP) framework and process tool provide an innovative solution to DIDRR implementation; it focuses on the preparedness of PwD in conjunction with emergency personnel. The P-CEP is the result of a co-design process involving various stakeholders, including PwD and their support networks. The Capability Approach is the pillar of the entire P-CEP concept, which interrelates factors that enable personal emergency preparedness and the principles of person-centered planning. In this way, the emergency management can evaluate the preparedness, capabilities, and the necessity of support for PwD and engage in the creation of local community-based Disability Inclusive Disaster Risk Reduction (DIDRR) along with them. The P-CEP adopts a comprehensive hazard strategy and enables self-evaluation and individual emergency preparedness planning for both natural disasters and other emergencies (for instance, a house fire or a pandemic). The P-CEP has three components: (a) a capability framework consisting of eight elements to help self-assessment of one's strengths and support needs; (b) principles that guide the collaborative efforts of all the stakeholders involved to make customized emergency preparedness planning possible; and (c) four process steps that facilitate the gradual increase of preparedness activities and, at the same time, help to establish connections between PwD, their support services, and emergency personnel and is also depicted in Fig. 7.1 (Villeneuve, 2022).

The meaningful involvement of disabled individuals in disaster planning, which is based on community requirements, is not only an assertion grounded on rights but also a necessity for building communities that are resilient practically. There is proof that the exclusion of handicapped individuals from preparedness activities

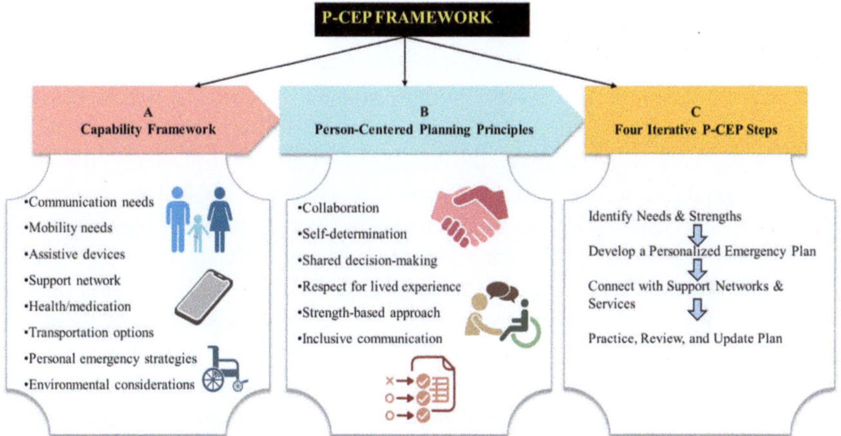

Fig. 7.1 Overview of the P-CEP Framework outlining capabilities, person-centered principles, and iterative steps for personalized emergency planning for persons with disabilities. This figure presents the Personalized-Capability Emergency Planning (P-CEP) Framework, which supports individuals with disabilities in developing tailored emergency preparedness plans. The framework integrates three key components: (A) a capability assessment that considers communication, mobility, assistive devices, health needs, support networks, transportation options, and environmental factors; (B) person-centered planning principles emphasizing collaboration, self-determination, shared decision-making, respect for lived experience, strength-based approaches, and inclusive communication; and (C) four iterative steps involving identifying needs and strengths, creating a personalized emergency plan, connecting with support networks and services, and regularly practicing, reviewing, and updating the plan. Together, these components help ensure inclusive, effective, and sustainable preparedness for persons with disabilities

has been raising their risk during emergencies and that this situation is rooted in historical worldwide social, structural, and institutional barriers. However, among the world's top frameworks like the Sendai Framework, UNCRPD, and P-CEP, as one of the new tools, there is a positive change in the way society sees persons with disabilities—recognizing them as co-partners in a process rather than aid-dependent and inactive recipients. The way to disability-inclusive disaster risk reduction is the honest collaboration of local authorities, emergency staff, disability support groups, and disabled persons themselves. By means of universal design, communication that is made accessible to everyone, and person-centered preparedness planning, communities will be able to break down the long-standing inequities and establish an environment where all people, regardless of their physical or mental capability, are educated, empowered, and supported as part of the disaster relief process: before, during, and after the disasters. The acknowledgment of the disabled in this context has to transition from being merely a mumbled promise to becoming an operational reality that always exists in all phases of emergency management.

7.3 Role of Disabled Persons' Organizations (DPOs)

Disability Persons' Organizations (DPOs) are groups primarily run by and for people with disabilities who advocate for their rights and offer services to the disabled community. DPOs were key players in the Asian and Pacific Decade of Disabled Persons (1993–2002) regarding its orientation and priorities. They were certainly present at the very beginning; for the Meeting to Launch the Decade, the Asia–Pacific Regional Council of Disabled Peoples' International ratified the importance of disability-led advocacy in the decision-making process. DPO engagement was also demonstrated by the significant presence of 12 major NGOs that were working on disability issues and were thus instrumental in policy dialogue, awareness-raising, and agenda setting. DPOs expressed through the NGO Symposium and their formal statements that the Meeting's report covered key points concerning full participation, equality, and rights-based approaches, thus making the Decade's Proclamation and Agenda for Action based on lived experiences. Through their cooperation with governments, UN agencies, and regional organizations, a multi-stakeholder commitment was created that would empower, make accessible, and promote the social inclusion of PwD throughout the Pacific region (Grills et al., 2020; UNESCAP, 1993).

Among the most important projects for knowledge management and capacity-building in the Pacific disabilities sector, one can mention the Innovation to Inclusion (i2i) Program. The program, which is coordinated by the Pacific Disability Forum and aided by various development partners, aims to make disability-inclusive development, thus systematically capturing, sharing, and applying information (Haque, 2025). The plan consists of various objectives, with the main one being to enhance disability-inclusive development in the Pacific region. First, it plans to pinpoint and share the best practices that will lead to the right policy and practice changes for inclusive approaches. Training not only upgrades the DPOs' capacity but also reinforces their ability to advocate for rights and deliver necessary services (Taylor, 2018). The peer learning and South-South cooperation among Pacific countries will surely lead to the sharing of experiences and solving problems together. The program will also be the one to improve the standard and reach of disability data and research, therefore, making the government's policy decisions based on facts rather than myths. Last but not least, it will always be the one to promote and support innovation not only in assistive technology but also in inclusive service delivery, thus ensuring the solutions are attuned to the needs of the PwD (Macanawai, 2009).

The i2i program is doing its best to make the knowledge application promoted through the Knowledge Application and Learning happen practically in the Pacific disabilities sector by means of a wide variety of strategies. One of these is giving technical support to governments in the area of disability policies and making sure that these policies are based on inclusive values. The other strategy is to train service providers, offering them the necessary skills for the smooth and effective application of inclusive practices. Mentoring ties are developed between experienced and emerging leaders within DPOs, enhancing advocacy and leadership capacity. Additionally, i2i supports pilot initiatives that test innovative ideas, with evaluation frameworks in place to quantify impact and inform future efforts. To promote disability-inclusive development across the region, policy discussion platforms bridge the gap between evidence and action by connecting research to decision-making (Haque, 2025).

Between 2020 and 2024, the i2i program achieved tremendous gains in increasing disability inclusion throughout 12 Pacific nations. Eight national disability policies were developed or revised, over 320 government officials, service providers, and DPO representatives were taught on inclusive approaches, and 47 case studies showcasing innovative practices were documented. In addition, 15 pilot initiatives were implemented, with 11 successfully scaled or duplicated in different contexts. A regionally accessible communications toolbox was also developed and delivered to more than 200 organizations. The leadership of people with disabilities through DPOs, which guaranteed authenticity and rights-based approaches, the early incorporation of accessibility into all knowledge products, the cultural adaptation of frameworks and resources to Pacific contexts, the involvement of a wide range of stakeholders, including governments, service providers, donors, and academia, and regional coordination that promoted economies of scale and cross-national learning, all contributed to these accomplishments. The initiative, however, eventually emerged with major challenges and limitations still to overcome. The reach of digital platforms was limited by the restrictions on internet connectivity on far-off islands, and the lack of proper baseline data made accurate measurement of the impact impossible. There were also doubts about the continuity of knowledge management systems due to the uncertainties regarding the long-term funding of the project. Language barriers persisted, as the majority of materials were provided in English, which made them less accessible to the local languages.

The initiative, however, managed to surmount such major problems and restrictions. Digital platforms were not able to reach the farthest islands due to poor internet connectivity, and the lack of baseline data hindered precise impact measurement. The lack of funding for the foreseeable future raised questions about the sustainability of the knowledge management systems set up. The language barrier was still an issue, as the materials were mostly in English, which made them less accessible to people with local languages (Haque, 2025).

7.4 Local Disability-Inclusive Initiatives: Lessons from the Pacific

The position of the Pacific area on the globe makes it prone to disasters quite often, and hence it is subject to heavy climate and disaster risks. Several projects in the various island countries across the ocean are showing the world no, rather than relying on modern technologies, they are using traditional knowledge and focusing on community resilience, including people with disabilities. The projects, in addition to all the above points, also give very important advice: the use of professional and culture-friendly techniques results in better preparedness and the so-called inclusivity during times of crisis.

7.4.1 Samoa: Commendable Resilience and Community-Based DRR Schemes

Samoa has put in place a village-based disaster risk reduction model that can markedly lessen the negative effects of global warming. The government, along with certain non-governmental organizations, has been working side by side with the local councils in revising the Climate Information Management plans and the relocation plans for the villages in the danger zones, which take into account the disaster risk as determined by the changing climate impacts. What is more, the disabled persons in the village councils are therefore present and active, which means they can communicate their views throughout the deliberation. The support from the faith-based and traditional leadership is the root cause of these initiatives as they attract the communities, formulate inclusive disaster protocols, and empower them. The training programs have already reached several hundreds of village leaders who are thus capable to properly plan for relocations and adapt to climate change through the means provided (ISDR, 2012; Toailoa, 2017; UNDRR, 2022).

7.4.2 Fiji: Disability-Inclusive Emergency Planning Frameworks

The disability-inclusive emergency planning frameworks in Fiji have been significantly boosted by the guidance of the Fiji Disabled Peoples Federation (FDPA). The FDPA works with government agencies, such as the Fiji Red Cross Society and the DISMAC (Disaster Management Authority), to include disability matters in disaster planning and response. This encompasses the creation of user-friendly evacuation shelters for people with disabilities, as well as the conduct of community drills that include participants with disabilities. Apart from this, the FDPA also supports training and advocacy to build support for disability-inclusive disaster policies at both the national and community levels. The actions they take ensure that mobility aids and other assistive devices are included in relief support during emergencies, reducing the limitations disabled persons face (FDPF, 2022; GPDRR, 2025; PDF, 2011).

7.4.3 Vanuatu: Customary Knowledge and Inclusive Community Preparedness

Vanuatu, to some extent, reflects the use of customary knowledge to make the whole community climate change disaster-resilient. Provincial and meteorological agencies acknowledge and support the use of early warning and preparedness strategies based on traditional warning signs and ancestral wisdom. The communities are taught to use traditional knowledge monitoring forms and tools, such as the ClimateWatch App, to collect and understand environmental data, which is how overcoming and scientific approaches are being implemented in the community. Likewise, the accessibility mapping programs of local communities have taken into account people with disabilities when mapping evacuation routes, thus making shelters and escape paths suitable and easy to use for all community members (SPREP, 2024; UNDRR, 2024).

7.4.4 Solomon Islands: Strengthening Grassroots Participation

The Solomon Islands devoted itself to involving the masses in disaster preparedness and response by including the disabled community in all aspects of the process, such as community hazard mapping and climate-induced disaster response planning. Locally, disability groups provide vital support by pointing out the shortcomings and bringing in disability-friendly solutions. Disability perspectives have been effectively integrated into the improvement of cooperation, early warning systems, and disaster information management through Disaster READY partnerships. Furthermore, the activities are in line with gender and protection considerations within community disaster management committees and therefore encourage a more holistic and inclusive approach to the treatment of climate-related emergencies (AHP, 2023; King et al., 2019).

7.5 Policy Implications for Disability-Inclusive Disaster Management

The countries making up the Pacific Islands find themselves vulnerable to the fury of nature in the form of cyclones, earthquakes, and rising sea levels; thus, it would be the best if the policy-makers were to include disability in disaster management plans. The lessons learned from local projects in Samoa, Fiji, Vanuatu, and the Solomon Islands have paved the way for new policy directions; specifically, the targeting of disability views in the national disaster risk reduction (DRR) strategies, the appointment of village councils for participatory planning, and the mixing of traditional knowledge with accessibility measures. These approaches not only ensure fairness but also enhance community ties as they help to alleviate the hardships of disabled people (approximately 15% of the population in the Pacific Islands), who are normally the most negatively impacted by the barriers during evacuation and relief activities. Standard techniques include incorporating considerations for people with disabilities into multi-hazard early warning systems and national plans, and making it compulsory for local committees to be inclusively trained. These regulations support justice by making accessibility standards mandatory in nations such as those in the Pacific Islands Forum, including for shelters and evacuation procedures (King et al., 2019; PDF, 2011).

The three Asian and Pacific Decades of PwD, along with the Incheon Strategy and the Beijing Declaration and Action Plan, demonstrate that the region is committed to disability inclusion grounded in rights, evidence, and a strong legal basis. The final assessment of the 2013–2022 Decade shows that although there have been advancements in law, data systems, and the involvement of different parties, there are still considerable gaps, especially in reproductive and sexual health, poverty alleviation, and DRR. Poverty and employment, political participation and decision-making, physical and digital accessibility, social protection and health care, early childhood intervention and education, gender equality and women's empowerment, disaster risk reduction, disability statistics, ratification and harmonization of the CRPD, and collaboration for disability-inclusive development are all areas where the Incheon Strategy's ten goals are being implemented, as shown in the table. The most recent information was obtained from secondary desk research, replies from international organizations operating in the area to the ESCAP short survey, and the results of the ESCAP government survey on the final evaluation of the implementation of the third decade. The examples for each nation were derived from government survey responses, unless otherwise noted. Government reporting on core indicators linked to DRR has been limited, and this not only reveals the need to improve but also that the gaps in providing disability- disaggregated datasets, inclusive early warning systems, and creating policies which involve OPDs in disaster planning, implementation, and evaluation processes are still huge (ESCAP, 2022; UN, 2006).

As the number of climate-related disasters in the Pacific skyrockets, the Jakarta Declaration (2022–2032) prioritizes disability-inclusive disaster management. The policy implications arising from this situation are the need to create a regional DRR framework that harmoniously connects the CRPD and SDGs; the establishment of OPD participatory rights in national disaster management; ensuring that evacuation routes and communication networks, as well as relief systems, are all accessible to people with disabilities; and finally, the increased fostering of cooperation among different sectors such as government, civil society, and humanitarian actors. The results from ESCAP's surveys among government and CSOs show that just policy reform will not be enough to give the required inclusive DRR; there needs to be capacity-building, funding commitments, and systematic data collection to ensure that people with disabilities will be fully protected and actively involved in the process of disaster management (preparedness and response) at all levels (ESCAP, 2012, 2018; UNDRR, 2015).

The preponderance of nations in the Pacific region have espoused laws on disaster risk reduction that include PwD, such as the national disaster management acts that require plans to address the various needs of the people, including those of people with disabilities. Data gathering on the most vulnerable groups and setting aside funds for the construction of infrastructure accessible to persons are among the policy provisions. Regional collaborations, such as those with the Pacific Community, advocate for uniform rules that guarantee participation in national committees. The governing principles allocate responsibility for disaster risk reduction to rural areas through local administrative systems, where inclusive discussions and risk assessments are obligatory. Non-governmental religious organizations are funded to develop public awareness campaigns, and at the same time, accessibility features become part of community-based early warning systems. It is usual to have annual drills and simulations to reinforce group readiness (AHP, 2023; GPDRR, 2025).

The majority of the nations and regions that participated in the ESCAP government survey reported having a national disability law. The participation of people with disabilities was noticeable throughout the enactment of the law, particularly in the planning and design phase, the implementation phase, and the monitoring and evaluation phase. Their involvement was not limited to, but included consultations, drafting committees and task forces, co-authoring implementation plans, assisting with service delivery, and being present in monitoring bodies, evaluation topics, and audit feedback. A large number of governments also implemented anti-discrimination measures, which were either rooted in disability laws or established through constitutional and sector-specific legislation, such as education and employment laws (ESCAP, 2022; Kanter, 2014).

Disability-inclusive development was promoted through national disability policies, strategies, or action plans in most of the countries surveyed. People with disabilities were involved throughout the policy cycle. They contributed during the planning and design phase by giving interviews, participating in workshops, attending public forums, holding meetings, serving on steering committees, and, in one case, co-authoring the action plan. They were also engaged in monitoring and evaluation by serving on evaluation committees, helping with assessments, and providing structured feedback, such as through questionnaires sent to organizations of persons with disabilities (Degener, 2016; Kanter, 2014; UNICEF, 2024).

7.6 Future Directions

The progress achieved so far under the Pacific Decade of PwD, the Incheon Strategy, and the Beijing Declaration and Action Plan is quite remarkable; however, PwD across the region still faces daily challenges that are hard to overcome. The Asia–Pacific region will fail to reach the majority of the Incheon Strategy targets on time, according to ESCAP's evaluation. To move from an ableist to a truly inclusive attitude, integrate disability perspectives across all sectors, and improve coordination across ministries in charge of gender, ageing, and disability issues, a whole-of-government and whole-of-society approach is required. The involvement of the private sector must be increased, regional cooperation must be improved, and the role of disability organizations must be strengthened. Participation should move from mere consultation to co-design and co-implementation, with sufficient funding and skilled staff, and with special attention to the most disadvantaged disabled groups (ESCAP, 2022).

The Jakarta Declaration (2022) boldly states that the world is a changed place and that the needs of PwD cannot be neglected. The document points out that the demographic revolution—decrease in birth rates and increase in longevity—along with the digital age, climate change, and pandemic recovery, are among the new challenges that must be met with the urgency that is the hallmark of disability-inclusive development. Governments are asked to accelerate their implementation of frameworks and strategies such as the Beijing Action Plan, the Incheon Strategy, the CRPD, and the 2030 Agenda. The main fields of action are: making national laws compliant with the CRPD; giving the disabled the opportunity to participate; abolishing barriers to accessibility by means of universal design; giving the private sector tax breaks for hiring people with disabilities; instituting gender-sensitive, life cycle-based policies; investing more in inclusive education, decent work, and welfare; and improving health, rehabilitation, and independent living services. Not only that, but also better disability data systems, harmonized indicators, and more active inclusion in SDG monitoring are extremely important (ESCAP, 2022).

Civil society organizations, comprising OPDs, are strongly recommended to invest in technical skills development, research and monitoring, intergenerational leadership, building knowledge hubs through strengthening umbrella OPDs, and forming partnerships with larger networks to further cross-sectoral disability inclusion.

7.7 Conclusion

The Pacific region has taken the lead in various ways to ensure that PwD are included in disaster preparedness, leveraging local innovations and traditional knowledge alongside existing regional policy frameworks. Through community-based initiatives in Samoa, Fiji, Vanuatu, and the Solomon Islands, it is demonstrated that the full participation of PwD not only can but also greatly does impact early warning systems, accessibility, evacuation planning, and overall resilience. However, according to the last review conducted by ESCAP, the region still lags in achieving key goals of the Incheon Strategy, especially in disaster risk reduction, accessibility, and disability-disaggregated data systems. Nonetheless, a multitude of challenges, such as the persistence of ableism, inadequate funding, poor coordination, and a lack of institutional mechanisms, continue to hamper the realization of inclusive DRR outcomes. The future of disability-inclusive disaster management should be based on co-design and co-responsibility, with active participation of OPDs in national and local DRR systems. The priorities that are essential for a country to have its national legislation aligned with the CRPD, and the universal design standards that have to be implemented, the investments that need to be made in inclusive education and social protection, and the data systems that need strengthening are the very ones mentioned in the previous sentence. Traditional knowledge that applies to disaster risk reduction will be used in conjunction with modern tools; there will be greater private sector involvement; and regional cooperation will lead to the survival of the fittest. The Jakarta Declaration (2022–2032) noted that disability inclusion should be a priority in tackling new challenges, including climate change, digital transformation, ageing, and post-pandemic recovery. Anyway, it is only through various, multi-sectoral, and culturally based strategies that Pacific communities will be able to guarantee the complete protection, empowerment, and meaningful inclusion of people with disabilities in every aspect, i.e., before, during, and after disasters.

References

Abbott, D., & Porter, S. (2013). Environmental hazard and disabled people: From vulnerable to expert to interconnected. *Disability & Society, 28*(6), 839–852. https://doi.org/10.1080/09687599.2013.802222

Australian Humanitarian Partnership. (2023). *Disaster READY Solomon Islands.* https://australiahumanitarianpartnership.org/disaster-ready-solomon-islands

Degener, T. (2016). Disability in a human rights context. *Laws, 5*(3), 35. https://doi.org/10.3390/laws5030035

Edmonds, C., & Noy, I. (2018). The economics of disaster risks and impacts in the Pacific. *Disaster Prevention and Management: An International Journal, 27*(5), 478–494. https://doi.org/10.1108/DPM-02-2018-0057

ESCAP, U. (2012). *Incheon strategy to "Make the Right Real" for persons with disabilities in Asia and the Pacific.* https://www.unescap.org/resources/incheon-strategy-%E2%80%9Cmake-right-real%E2%80%9D-persons-disabilities-asia-and-pacific

ESCAP, U. (2018). *Incheon Strategy to "Make the Right Real" for persons with disabilities in Asia and the Pacific and Beijing declaration including the action plan to accelerate the implementation of the Incheon strategy.*

ESCAP, U. (2022). *A three-decade journey towards inclusion: assessing the state of disability-inclusive development in Asia and the Pacific.* https://www.unescap.org/kp/2022/three-decade-journey-towards-inclusion-assessing-state-disability-inclusive-development

Fiji Disabled Peoples Federation. (2022). *Fiji disability-inclusive community-based disaster risk management toolkit.* https://pacificdisability.org/wp-content/uploads/2022/01/Fiji-Disability-Inclusive-Community-Based-Disaster-Risk-Management-Tool.pdf

Gaillard, J. C., & Mercer, J. (2013). From knowledge to action: Bridging gaps in disaster risk reduction. *Progress in Human Geography, 37*(1), 93–114. https://doi.org/10.1177/0309132512446717

Global Platform for Disaster Risk Reduction. (2025). *Building inclusive disaster preparedness through community-based early warning and response systems: lessons learned from Bangladesh and Fiji.* https://globalplatform.undrr.org/conference-event/building-inclusive-disaster-preparedness-through-community-based-early-warning

Grills, N. J., Hoq, M., Wong, C. P. P., Allagh, K., Singh, L., Soji, F., & Murthy, G. V. S. (2020). Disabled People's Organisations increase access to services and improve well-being: Evidence from a cluster randomized trial in North India. *BMC Public Health, 20*(1), 145. https://doi.org/10.1186/s12889-020-8192-0

Haque, A. (2025). *Disability-inclusive communications and knowledge management: A pressing need in Pacific Island Countries.* Available at SSRN 5736924.

Hay, J. E., & Mimura, N. (2013). Vulnerability, risk and adaptation assessment methods in the Pacific Islands Region: Past approaches, and considerations for the future. *Sustainability Science, 8*(3), 391–405. https://doi.org/10.1007/s11625-013-0211-y

Hemingway, L., & Priestley, M. (2006). Natural hazards, human vulnerability and disabling societies: A disaster for disabled people?. *Review of Disability Studies: An International Journal, 2*(3).

Hewitt, K. (Ed.). (2019). *Interpretations of calamity: From the viewpoint of human ecology.* Routledge. https://doi.org/10.4324/9780429329579

ISDR, U. (2012). *Towards a post-2015 framework for disaster risk reduction.*

Jackson, G., McNamara, K., & Witt, B. (2017). A framework for disaster vulnerability in a small island in the Southwest Pacific: A case study of Emae Island, Vanuatu. *International Journal of Disaster Risk Science, 8*(4), 358–373. https://doi.org/10.1007/s13753-017-0145-6

Kailes, J. I., & Enders, A. (2007). Moving beyond "special needs" A function-based framework for emergency management and planning. *Journal of Disability Policy Studies, 17*(4), 230–237. https://doi.org/10.1177/10442073070170040601

Kanter, A. S. (2014). *The development of disability rights under international law: From charity to human rights.* Routledge. https://doi.org/10.13140/2.1.3039.6484

Kates, R. W. (1971). Natural hazard in human ecological perspective: Hypotheses and models. *Economic Geography, 47*(3), 438–451. https://doi.org/10.2307/142820

King, J., Edwards, N., Watling, H., & Hair, S. (2019). Barriers to disability-inclusive disaster management in the Solomon Islands: Perspectives of people with disability. *International Journal of Disaster Risk Reduction, 34*, 459–466. https://doi.org/10.1016/j.ijdrr.2018.12.017

Lavell, A., & Maskrey, A. (2014). The future of disaster risk management. *Environmental Hazards, 13*(4), 267–280. https://doi.org/10.1080/17477891.2014.935282

Macanawai, S. S. (2009). Disability in the Pacific. *Inclusive Education in the Pacific*, 56–69.

Nations, U. (2006). *Convention on the Rights of Persons with Disabilities (CRPD), U.* Division for Social Policy and Development Disability, 1–31. https://social.desa.un.org/issues/disability/crpd/convention-on-the-rights-of-persons-with-disabilities-crpd

O'keefe, P., Westgate, K., & Wisner, B. (1976). Taking the naturalness out of natural disasters. *Nature, 260*(5552), 566–567. https://doi.org/10.1038/260566a0

Pacific Disability Forum. (2011). *Disability inclusiveness in disaster and risk reduction Mgt in Fiji, Progress Report to AUSAID.* https://www.dfat.gov.au/sites/default/files/disaster-preparedness-monitoring-disability.pdf

Quarantelli, E. L. (1987). Disaster studies: An analysis of the social historical factors affecting the development of research in the area. *International Journal of Mass Emergencies & Disasters, 5*(3), 285–310. https://doi.org/10.1177/028072708700500306

Robinson, A., & Kani, S. (2014). Disability-inclusive DRR: information, risk and practical-action. In *Civil society organization and disaster risk reduction: The Asian dilemma* (pp. 219–236). Springer Japan. https://doi.org/10.1007/978-4-431-54877-5_12

Secretariat of the Pacific Regional Environment Programme. (2024). *Vanuatu empowers communities to harness traditional knowledge for climate resilience.* https://www.sprep.org/news/vanuatu-empowers-communities-to-harness-traditional-knowledge-for-climate-resilience

Stough, L. M., & Kang, D. (2015). The Sendai framework for disaster risk reduction and persons with disabilities. *International Journal of Disaster Risk Science, 6*(2), 140–149. https://doi.org/10.1007/s13753-015-0051-8

Taylor, M. D. (2018). *Pacific Islands forum leaders 2018 decisions on sustainable development.* Pacific Islands Forum Secretariat.

Toailoa, A. S. (2017). *Exploring community engagement in climate change planning: The case in Samoa* (Doctoral dissertation). Open Access Te Herenga Waka-Victoria University of Wellington. https://doi.org/10.26686/wgtn.17065382

UN.ESCAP. (1993). *Asian and Pacific decade of disabled persons, 1993–2002: The starting point.* https://hdl.handle.net/20.500.12870/4381

UNDRR. (2022). *Disaster risk reduction in Samoa: Country update 2022, United Nations Office for Disaster Risk Reduction (UNDRR), Sub-Regional Office for the Pacific.* https://www.undrr.org/publication/disaster-risk-reduction-samoa-status-report-2022

UNDRR. (2015). *Sendai framework for disaster risk reduction 2015–2030.* https://www.undrr.org/publication/sendai-framework-disaster-risk-reduction-2015-2030

UNDRR. (2024). *Vanuatu empowers communities to harness traditional knowledge for climate resilience.* https://www.preventionweb.net/news/vanuatu-empowers-communities-harness-traditional-knowledge-climate-resilience

UNICEF. (2024). *An inclusive world starts with me, with you, with all of us.* https://www.unicef.org/media/134511/file/An%20inclusive%20world,%20starts%20with%20me,%20with%20you,%20with%20all%20of%20us.pdf

UNISDR. (2013). *Living with disability and disasters UNISDR 2013 survey on living with disabilities and disasters—Key findings.* https://www.unisdr.org/2014/iddr/documents/2013DisabilitySurveryReport_030714.pdf

Villeneuve, M. (2022). Disability-Inclusive emergency planning: Person-Centered emergency preparedness. *In Oxford Research Encyclopedia of Global Public Health.* https://doi.org/10.1093/acrefore/9780190632366.013.343

Wisner, B., & Wisner, B. (2004). *At risk: Natural hazards, people's vulnerability and disasters.* Psychology Press.

Wisner, B., Gaillard, J. C., & Kelman, I. (2012). *Handbook of hazards and disaster risk reduction.* Routledge.

Chapter 8
Advocacy, Policy, and the Future of Inclusive Resilience

Abstract The chapter thoroughly examines advocacy, inclusive policy, and technology as the major pillars that protect the disabled community in the Asia–Pacific region from disasters. Despite the economic growth, people with disabilities remain the most affected group. They have very limited mobility when it comes to public transportation and their access to jobs, healthcare, and disaster risk management is no better. The chapter mentions international, regional, and national standards, including the CRPD and the SDGs, that together form a legal shield, while also drawing attention to the little progress made in the area of policy application. In addition, it goes further to explain how the global mega forces of climate change, digital transformation, AI, and governance are intermingling with the concept of disability inclusion. The Pacific islanders' fast-moving technology has both advantages and disadvantages; thus, it is even more critical to make sure that there are accessible ICT, culturally based innovations, more robust interagency collaboration, and community partnerships that last. The chapter is well-grounded in the use of case studies and regional assessments, from which it builds the argument for the role of inclusive practices, effective participation, and supportive ecosystems in the realization of equitable resilience. To conclude, by identifying areas for future research, society will be able to provide for the needs of people with disabilities in terms of accessibility, digital inclusion, and community adaptation strategies.

Keywords Advocacy · ICT · Digital inclusion · Assistive technologies · Intersectionality

8.1 Introduction

The impact of disasters on society is huge and of many different kinds, and one of the main effects is that marginalized groups become the most vulnerable and extinction of their rights gets to be more intense and social injustices that are already there get to be more escalated. People who are already economically and socially disadvantaged are during crises pushed more to the periphery, thus making them the ones who suffer

S. M. Thomas and R. Veerabathiran, *Disability, Disaster, and Resilience in Oceania*,
SpringerBriefs in Modern Perspectives on Disability Research,
https://doi.org/10.1007/978-981-95-8535-9_8

the most from the bad consequences of disasters. Such a population is often excluded from the decision-making process, is not able to affect policy development to any significant extent and is facing a lot of obstacles when trying to gain access to the opportunities that would elevate their living standards. As a result, individuals face systematic discrimination in several fields, including public administration, healthcare, and education, which further isolates them from mainstream social institutions. People with disabilities are among the most vulnerable groups and are particularly vulnerable during disasters (Jevtić et al., 2025; Joseph & Doon, 2023; Mendis et al., 2023).

During emergencies, the rights of people with disabilities, unlike those of other groups, are protected by a strong legal background at both the global and local levels, which has already been established. The ICCPR, the ICESCR, and the CRPD, plus their Optional Protocols, are the main international treaties. The CRPD, among them, stands out for being the very first international convention with legally binding power that is exclusively addressed to the PwD; therefore, it is recognized as a significant milestone because the attention in this case is not only on the mere existence of rights and freedoms but on the actual enjoyment of them through the authorities of the states concerned (Schulze, 2010). The UN 2030 Agenda for Sustainable Development incorporates disability rights into several universal guidelines sectors, such as (a) economic development (b) penury alleviation, (c) gender equity, and (d) global access to basic education, in addition to international conventions. To build more resilient and equitable communities, the Sustainable Development Goals (SDGs) stress the significance of involving people with disabilities in (a) data gathering activities, (b) workforce development, and (c) sustained catastrophe risk reduction plans (Colglazier, 2015).

Numerous legislative and strategic frameworks at the regional level promote policies for the security and blending of persons with disabilities. The rights of humans in Europe (ECHR) and Protocol No. 12 (2000) laid down the basis for the abolition of discrimination; the European Action Plan for PwD (2006–2015) focused on social inclusion, and the Council of Europe's Disability Strategy (2017–2023) identified five key issues: (a) equity and fairness; (b) awareness-raising; (c) convenience; (d) equal acknowledgement before the law; and (e) safeguard from mistreatment, violence, and disregard. The EU Disability Strategy 2010–2020: A Barrier-Free Europe stresses the necessity of emergency preparedness planning that is sensitive to the special needs of people with disabilities and helps to eliminate social barriers that hinder the full participation of people (Cuenca, 2012; Priestley et al., 2016; Schokkenbroek, 2004).

Inclusion does not just suggest the right—it is a basic human right indeed. To ensure that everyone has equal access to safety measures, emergency services, and legal protection is the first step in lowering the total disaster risk. Besides, it is very important to the building of a more robust and flexible society; so the dismantling of the barriers that keep the disabled away from DRR programs would be a gradual process. A resilient community is one that no one is excluded from and that the empowerment of the disabled is through the acknowledgment of their specific needs, new ideas, and essential input. The risks are not eliminated completely, but the involvement of all will lead the community to find ways of being prepared, therefore reducing vulnerability and increasing the overall flexibility to disasters (Jevtić et al., 2025; Sagala et al., 2024; Spurway & Griffiths, 2016).

The Pacific region is rapidly changing its electronic western media reception, also including satellite TV and internet cables cutting through the ocean. Improved digital connectivity implies that the availability of information and communication technologies (ICT) will be ever more essential for Pacific people to take part in worldwide systems. ICT has played a vital role in the support, maintenance, and furtherance of the interaction between people with disabilities and their non-disabled counterparts (Ratliffe et al., 2012).

The chapter interrelates the legal, policy, and advocacy factors, which are the major determinants of the future of disaster resilience in the context of disability inclusion. It elaborates on the state obligations specified by international and regional human rights frameworks, on the persistent social attitudes and the structural barriers that prevent the complete integration of persons with disabilities, and on the potential of ICT and emerging technologies to increase avenues for inclusion particularly in the Pacific area. In addition to what has been mentioned above, the chapter highlights the crucial input of DPOs, vibrant community leadership, and joint governance in reshaping the whole DRR process. By employing policy analysis combined with real-life case studies and strategies looking ahead, this chapter has not only laid the foundation for rethinking advocacy but also has bolstered inclusive policy making and the creation of a resilient and just future for persons with disabilities.

8.2 Reimagining Policy: Leadership of the Disabled Communities

The Pacific region's economic growth and development will be supported not only by the moral or humanitarian requirement but also by the inclusion of people with disabilities. As countries deal with rapid changes in the environment, technology, and demographics, the holistic method must be applied to every facet of economic planning and policy-making. Besides, the consideration of people with disabilities will not only slow down social development but also diminish effective retorts to the megatrends determining the future of the region (Tjondronegoro et al., 2025).

Persons with disabilities are among the poorest and most marginalized people of the world, which is, in fact, a significant majority; they are the ones who experienced the COVID-19 pandemic as a health threat with greater intensity and are increasingly susceptible to the negative impacts of the ongoing climate crisis that is getting worse constantly. The Pacific area is inhabited by about 700 million disabled individuals, almost one-sixth of the people of the region; thus, disability should be an important future issue waiting to be solved in the coming years. Economic progress and growth in the Pacific area are necessary for the long-standing financial strength and wealth of the whole population, especially persons with disabilities, as they create job opportunities, professional and social facilities, improved transportation and accessibility, and easier access to assistive tech. However, active integration of disability inclusion into policy and business strategies is essential if these gains are to be realized. Long-term patterns (megatrends or five mentionable ones) that will shape the region's future economic, technology, and social systems will define the direction of the Pacific Islands. By the very same long-term patterns, the region's policymakers and businesses can build their responses to emerging opportunities, challenges, and sustainable development that will result from this understanding. Pacific will be open or closed regarding the above-mentioned trends in cases of discrimination against disabled people. As the Pacific states constantly and at an increasing rate face demographic, high-tech, and ecological changes, an inclusive future will have to accompany and be integrated into every one of these changes. Inclusion will not be a minor issue but a gigantic macroforce that either hastens or hinders the social and economic progress of the entire region. Consequently, since it is itself a macroforce, nothing is more vital to the region's future than inclusion, more so than any of the identified megatrends (ESCAP, 2022b, 2023; Rechner et al., 2023; Tjondronegoro et al., 2025).

Pacific has witnessed a tremendous increase in the production of goods and services, but there are still gross inequalities, particularly for persons with disabilities, as the rich get richer and even the poorer become poorer through the divisions of classes, regions, and people. In rural and low-governance areas, where accessibility to basic needs is lacking, people with disabilities are cut off from participation in society due to their reliance on and unavailability of assistance in the form of technologies, education, healthcare, and employment. On top of this, disasters, pandemics, the prevalence of discriminatory attitudes, and the poor state of infrastructure have caused the situation to worsen. At the same time, many countries are moving towards adopting disability-friendly legislation and signing onto the CRPD. Yet, the story of implementation remains one of gaps that lead to weak enforcement and inconsistent access to technology, both digital and assistive. The benefits of disability inclusion extend not only to equity but also to the economic and social development of the region; inclusive societies can boost productivity, innovation, and resilience, especially as ageing populations and persons with disabilities rise. When disability inclusion is viewed as a strategic investment rather than a financial burden, it becomes easier for policymakers to cut down on long-term social and healthcare costs, tap into the range of skills that people have—including the ones of neurodivergent individuals—and to usher in a sustainable and inclusive growth that spreads across the region (Bharti, 2024; ESCAP, 2018; GFDRR, 2017; OECD, 2010; Tjondronegoro et al., 2025; Twigg et al., 2018). The various regional and international frameworks that support the disabled persons in the Pacific region are illustrated in Table 8.1.

Table 8.1 International and regional frameworks for the disabled persons

Framework	Scope	Key elements	Role in disability inclusion
ICCPR & ICESCR	Global	Protection of rights during emergencies	Establish a legal foundation for the rights of people with disabilities
CRPD	Global	First legally binding convention dedicated to PwD	Ensures enjoyment of rights, mandates inclusive DRR
SDGs (2030 Agenda)	Global	Economic development, poverty alleviation, gender equality, and education	Integrates disability rights into universal development efforts
European Convention on Human Rights (ECHR)	Regional (Europe)	Protocol No. 12 prohibits discrimination	Supports equal recognition and protection
European Action Plan for PwD (2006–2015)	Regional	Social inclusion	Strengthens policy for accessibility
Council of Europe Disability Strategy (2017–2023)	Regional	Equity, awareness-raising, accessibility, equal recognition, protection from violence	Guides rights-based inclusion
EU Disability Strategy 2010–2020	Regional	Barrier-free Europe	Includes disability-sensitive emergency planning
Sendai Framework for DRR	Global	Participation of PwD, resilience-building	Calls for OPD leadership in DRR and better disability data
Incheon Strategy	Asia-Pacific	CRPD implementation, data, multi-stakeholder participation	Core basis for Pacific disability policies
Pacific Framework for the Rights of PwD	Pacific Region	CBID, rights-based inclusion, SDG alignment	Drives regional disability policy and OPD involvement

8.2.1 Megatrends: Its Role

A megatrend denotes a set of detectable patterns in the economy, society, or even the environment that are, to some extent, predictable and will significantly alter life over the next few years. Six essential megatrend themes have been recognized for the period until 2040: (1) weather and power, (2) well-being, (3) geopolitics, (4) digital conversion, (5) AI and automation, and (6) authority, which are the reflections of the major trends identified through five important global trend reports over the last 10 years. Each trend is influencing the future in different ways—sustainability imperatives and ageing populations, power shifts among nations, growing digital networks, disruptions to the job market caused by new technologies, and new approaches among institutions are among the areas the trends are affecting. To cope with these trends, an inclusive approach is needed, which will guarantee a sustainable and fair future for the Pacific region (CSIRO, 2022).

Therefore, without explicit emphasis on inclusivity—especially about disability, policy measures stand to widen existing inequality against those who combine sex with poverty and ethnicity or location. For instance, women with disabilities face heightened risks of violence during crises, and people with disabilities living in poverty face increased economic insecurity and limited access to services. Hence, the concept of inclusion becomes the largest macroforce that can, hopefully, shape institutional, economic, and societal landscapes over the longer term, shaping how the region can leverage megatrends. While the coverage of equity and inequality below addressed major global trends, disability was rarely brought into the picture, despite the fact that inclusion is the essence of all megatrends (Tjondronegoro et al., 2025).

Disaster, inaccessible infrastructure, or sustainability measures failing to accommodate their need are only some of those dimensions whereby people with disabilities suffer disparities in climate and energy transition. Persistent health inequalities due to chronic illnesses, inaccessible services, and discrimination, primarily directed against women with disabilities and rural populations, are on the increase. People are further blocked from receiving essential care, safety, and assistance. The AI revolution represents a new opportunity to provide assistive technologies and remote care that can easily overcome accessibility issues. However, uneven infrastructure, insufficient affordability, and lack of accessibility to technology block the way for a large number of individuals. While AI and automation may necessitate individualized living experiences, the potential for perpetuating discrimination remains a concern, given biased datasets and weak regulations. Lastly, the area's governance systems range widely, with some inclined towards inclusive policies and others entangled in well-entrenched ableism and systemic barriers that regularize oppression (Nguyen-Trung et al., 2025; Tjondronegoro et al., 2025).

8.2.2 *Strengthening: Disabled Persons' Organization and Right-Based Frameworks*

DPOs/OPDs continue to be at the forefront of promoting Pacific inclusion by advocating for CRPD implementation and bringing together diverse community efforts. The Disabled Peoples Federation of Fiji has over 21 branches, where disabled leaders tactically assess accessibility and become part of village committees, thus influencing DRR and livelihoods. In Samoa, Loto Taumafai Society connects all CBID matrix components and continues to provide lifespan services, even amid funding gaps, alongside the national OPD Nuanua O Le Alofa. To strengthen this, government funding and training are needed, which was the case in the Cook Islands, where OPDs, after clarifying roles during COVID, thus encouraged shared responsibility (PDF, 2025; Sharma et al., 2018; UNDRR, 2023).

The CRPD and the Incheon Strategy serve as the foundation for the Pacific's disability policies and obligate the inclusion of disabled people through various means, such as data collection and multi-stakeholder involvement. The Pacific Framework for the Rights of PwD incorporates CBID as a prerequisite, thereby linking it to the SDGs and the 2050 Blue Pacific Strategy. The UNDRR's Sendai Framework calls on Pacific countries to allocate funds to support the leadership of OPD in Disaster Risk Reduction to address problems related to data on persons with disabilities and the construction of resilient infrastructure. Such frameworks are a transition from charity to rights, thus guaranteeing non-discrimination and access (ESCAP, 2022c; PDF, 2025; UNDRR, 2023).

Governance transformation shifts DRR power from the central to the community level through CBID, thereby incorporating people with disabilities into national plans. The Ministry of Health in the Solomon Islands is working together with 18 ministries to ensure the participation of people with disabilities at the ward level, and is also collaborating with NGOs such as Save the Children. The CBID group in Palau is advancing the humanitarian agenda by integrating climate components into the matrix of early warning and resilient livelihoods. UNDRR says that multi-stakeholder approaches encourage the OPD to take the lead in local decision-making and provide new budgets for the resilience of the Pacific region (PDF, 2025; UNDRR, 2023).

Intersectionality examines the different dimensions of disability, gender, and indigenous and cultural issues in Pacific policy. Men and women with disabilities are experiencing almost the same situation when it comes to violence, thus UNFPA is conducting assessments for access to SRHR in Fiji, Samoa, and Vanuatu for women with disabilities. In Papua New Guinea, disabled people's organizations are bringing gender, disability, and indigenous rights together through Talanoa dialogues, encouraging movements across. According to the NIDWAN report on ethnic violence, disabled indigenous women, who are often overlooked, are calling for integrated approaches to DRR (Cooms et al., 2024; Gurung, 2022; Kayess et al., 2014; Thongkuay, 2024).

8.3 Inclusion: Innovation, Accessible Technologies, and Future Research Agendas

The Pacific region is at a critical point in the debate about disaster resilience and disability rights, where the use of new technologies, access for all, and social equity dominate the discussion. Moreover, the small developing island nations are experiencing the worst consequences of climate change, cutting off their communication with the world and leaving them with little or no infrastructure. On the contrary, they are becoming the basis for creative, community-oriented solutions. New technologies' arrival, like the setting up of early warning systems for the disabled, mobile health platforms, and the manufacturing of assistive devices suitable for low-resource environments has allowed people to join in and be protected. Moreover, the Pacific island scientists are not only moving their attention but also making the quiet ones heard, providing the handicapped with the opportunity to express their opinions through the apparatus and regulations being produced. The power shift is not only about having the latest technology at your disposal; it is also around being culturally backed, relevant to the local situation, and fair. The West has always had the "superior" technology but this section insists on the unity of modern and inclusive research as a promising contributor not only to the future resilience of the Pacific Islands but also to their responsiveness to the demands of disabled people.

The development of a comprehensive ICT policy is influenced by the topography, history, principles, and governments of the U.S.-connected Pacific area. Under the auspices of the UN and other cooperation programs, improvement is significant as countries unite to create operative plans, enabling them to provide these technologies to the people who need them. The Pacific region, which is associated with the US, has different aspects such as geography, history, culture, and politics that are impacting the creation of a holistic ICT policy. When the UN and other joint efforts lead the way, nations cooperating to develop the right policies are already a big step forward towards making these technologies available and accessible to the targeted people. The University of Hawaii Center on Disability Studies (previously known as the University Affiliated Program for Persons with Developmental Disabilities) took the lead among the region's pioneers in implementing the Technology-Related Assistance for Persons with Disabilities (Tech Act) and the Americans with Disabilities Act (ADA) starting in the early 1990s. Through the combination of both civil rights movements, self-advocacy, self-determination, and person-centered planning, values were promoted, all of which are relatively new ideas to the Pacific region. It was the Tech Act and the ADA that legally defined and mandated the term "assistive technology." Assistive technology was defined as any gadget that helps a person with a disability feel more included at home, at school, in the workplace, and in the community. This encompassed both high-tech and low-tech, homemade and off-the-shelf products. Services like training, evaluation, and repair were also included in the definition. From 1997 to 2004, assistive technology was employed as a related service under the Individuals with Disabilities Education Act (IDEA) to help children with disabilities benefit from their education. When support plans were being created for challenged people and their families, disabling conditions were the first to show that technology is assistive. This applied to all age groups, including early intervention, special education, and vocational rehabilitation. These laws created a situation in which jurisdictions that received U.S. government funds for education, vocational rehabilitation, and other mandated services would have to treat their disabled citizens in accordance with the law (Butcher, 2010; Labelle, 2005; Park et al., 2021; Ratliffe et al., 2012).

8.3.1 The Biwako Millennium Framework

The Biwako Millennium Framework (BMF) of the UN Economic and Social Commission for Asia and the Pacific (UNESCAP, 2002) serves as the basis for policy recommendations to build a rights-based society for people with disabilities in the Pacific. The framework was accepted at an intergovernmental meeting of UNESCO (United Nations Educational, Scientific and Cultural Organization) that concluded the Decade of Disabled Persons in Japan in 2002, and it also extended these countries' commitment to creating a rights-based society for another 10 years.

The BMF has proposed seven priority areas for achieving a society in which people with disabilities can participate fully. One of these areas is "access to information and communications, including information, communication, and assistive technologies." The BMF is urging governments and agencies in the region to help provide access to information and communications technologies for people with disabilities by taking into account issues of accessibility, policies, information dissemination, and training.

The BMF is referred to as the foundation for their work by different organizations focusing on inclusive education and disability in the Pacific Islands. The Pacific Disabilities Forum's strategic plan (2007) states that young governments and small economies in the Pacific region are unable to provide the necessary support and services for the disabled population. Even though this strategic plan does not address ICT directly, it discusses the needs of the disabled population that must be addressed before access to ICT. There are three main areas of need: policy and inclusive planning development, community awareness, and information dissemination, which are the most pressing for the Pacific region. The project on Pacific Regional Initiatives for the Delivery of Basic Education (PRIDE), which involves a group of aid agencies and service providers based at the University of the South Pacific in Fiji, also cites the BMF as a justification for its regional inclusive education work (Puamau & Pene, 2009). Although they acknowledge the need to develop regional and national policies for assistive technology, their focus is on the more basic areas of raising disability awareness, training, and sharing information about disabilities. In some communities, disabilities are associated with negative traits like fear and shame. There will be no significant progress in the use of assistive devices to facilitate inclusion in schools, workplaces, and communities until these attitudes towards disability change.

Borg et al. (2009) noted that discussions of the CPRD emphasized the role of assistive technology in providing people with disabilities access to all human rights and urged national governments to develop and enforce equal policies and laws on assistive technologies. The CPRD has given the less-resourced nations time to gradually achieve the goals set forth, recognizing the difficulty that they will have in carrying out these policies. The authors stated, "such impediments should not be used as an excuse to ignore the provision of assistive technology in view of the benefit to human development" (Borg et al., 2009).

8.3.2 Digital Innovation: A Way Towards a Better Tomorrow

Along with the digitalization of products and services, the nurturing of digital transformation is changing the entire value-creation, management, use, and distribution processes with the help of disruptive technologies such as mobile connectivity, big data, and artificial intelligence (AI). Digital technologies and innovations offer a wide range of opportunities, including better access to information, education, skills training, employment, business opportunities, social benefits, and health and financial services. Nevertheless, without proper planning, implementation, and monitoring, digital technologies can worsen the situation of the disadvantaged. Also, the new barriers created will be specifically targeted at them, thus making it hard for them to get involved in the social, political, and cultural aspects of life. On the other hand, digital inclusion has been seen as "the right of everybody, everywhere, to have equal, meaningful and safe access to digital technologies, services and opportunities and to be allowed to use, lead and design them" (ESCAP, 2022d; Heaslip & Holley, 2023; Zian & Oguzhan, 2023).

It is indispensable to leverage digital innovations for inclusive and sustainable development. International and regional normative frameworks have repeatedly highlighted the pressing need to harness technology and innovation to improve the lives of marginalized people. The digital space is barred to people for several reasons, including, but not limited to, their gender, age, disability, geographic location, ethnicity, migration status, education, and socio-economic status. Thus, the digital rights of the most vulnerable people demand a thorough study of the barriers they face, as well as the power relations involved that overlap at different levels (Zian & Oguzhan, 2023).

The International Telecommunication Union (ITU) Smart Villages and Smart Islands initiative is also focused on uplifting the most isolated and deprived communities through improved broadband connectivity and affordability, the development of digital skills, and the provision of digital services. One of the flagship projects under the Smart Villages initiative in Pakistan, supported by the Ministry of IT and Telecom, social impact company TeleTaleem, the one-stop telemedicine provider Sehat Kahani, and other partners, has been successful in bringing digital education and health services to the people living in Gokina, a remote mountain village. In Vanuatu, the Government, together with the ITU, has commenced a pilot project in the southern Malekula region to bring about digital transformation at the community level. The inhabitants can now access the Internet via the Very Small Aperture Terminal (VSAT) infrastructure. Moreover, the project aimed to accelerate the digital literacy training it offers to the community, including the youth, women, people with disabilities, and others. The combined efforts have turned 20 isolated villages and islands into digitally-enabled communities (Heap & Hirmer, 2020; Umar, 2018; Zian & Oguzhan, 2023).

Giving underprivileged groups the chance to improve their digital skills to a level that makes them eligible to pursue STEM careers is a necessity. The need for a diversified, STEM-literate, and multidisciplinary workforce is widely accepted, as it is the only way to address current development challenges. The issue of the marginalization of certain groups in the STEM arena—frequently due to stereotypes and social currents—cripples the tech industry's capability to respond to various needs effectively. At an early age, girls and boys usually have similar levels of digital literacy, but as girls go through school, especially, they become less likely to acquire advanced digital skills and thus, fall behind. The situation globally is that women's participation in STEM-related higher education is a mere 30%; this is the case where the lowest enrollment of females is recorded for ICT, engineering, manufacturing, and construction, and natural science, mathematics, and statistics. Furthermore, globally, women occupy less than a quarter of the positions in science, engineering, and

ICT, and they are only 17% of the total inventors involved in international patents. Other marginalized populations in education and the job market often receive the same treatment in STEM-related fields (Chavatzia, 2017; ESCAP, 2022a; Ng, 2019; Women, 2022).

8.3.3 Future Perspectives

Frameworks, such as the Biwako Millennium Framework and others that build on it, underline the importance of ensuring that people with disabilities have the right to access ICTs and also push governments to create policies, disseminate information, and provide training to facilitate access to these technologies (Ratliffe et al., 2012).

The gradually recovering countries from the COVID-19 pandemic are steering towards more resilient, inclusive, and sustainable societies. Hence, closing down the digital divides and promoting digital inclusion have been placed on the agenda as the most important and urgent actions to be taken. Digital technologies have their pros and cons, which in turn call for the establishment of a future that is open, safe, and human-centered through constant actions that are in line with the UN Charter, the Universal Declaration of Human Rights, and the 2030 Agenda. Meaningful connectivity involves not only increasing the digital accessibility but also enhancing the quality, affordability, and usability of the services, thus, empowering the digital skills of the marginalized groups through support in STEM fields. Furthermore, it is necessary that the digital ecosystems keep on providing public employment services that are accessible, inclusive digital labor platforms, large support for MSMEs, and very strong social protection systems. A rights-based approach is indispensable in tackling the problems associated with technology and online violence; the safeguarding of rights such as privacy, freedom of expression, and data protection, etc. will be ensured (ESCAP, 2023).

The application of systematic measurement and monitoring is mandatory for the promotion of digital inclusiveness. The Government, on the other hand, needs to have the best possible data that is detailed according to various categories like gender, age, disability, religion, social class, etc., so that their responses can be more precise. The collaboration among different sectors and stakeholders is an absolute necessity for involvement that brings together governments, civil society (including organizations of marginalized groups), private sector players, academia, and aid agencies. If governments develop policies in partnership with people in vulnerable situations, they will not only become better acquainted with the specific contextual needs but also be able to develop quick and creative solutions. Stakeholders in the region should: (1) collect group-based data and utilize it for policy-making that corresponds to the needs; (2) ensure meaningful participation and lead for marginalized groups; (3) regularly exchange experiences and share best practices in digital inclusion; and (4) foster cooperation at regional and sub-regional levels for the improvement of capacity-building, financing, investment, and technology transfer (ESCAP, 2023).

8.4 Strengthening the Advocacy Ecosystems in the Pacific

By forming alliances with governments, academia, and the NGO sector, Pacific disability advocacy movements gain a stronger voice. They do this by sharing resources and expertise to build inclusive resilience. The Pacific Framework for the Rights of PwD (PFRPD) obliges the carrying out of consultations with disabled people's organizations (DPOs), governments, and stakeholders through the Informal Working Group on Disability (IWGD), allowing the building of multi-sectoral partnerships for actions aligned with the CRPD. In Fiji and Papua New Guinea, for instance, the Fiji Disabled Peoples Federation and PNG Assembly of Disabled People, as DPOs, work hand in hand with the governments on laws such as the Rights of PwD Act; this way, the civil society's contribution is taken into account during national consultations. Research partnerships are one of the ways academia is contributing; for example, in the Solomon Islands, where DPOs and universities are advocating for the Washington Group Questions in censuses, this is facilitating data-driven advocacy. These coalitions are also present in the DRR area, where the Pacific Disability Forum is reaching out to marginalized people with disabilities (OPD) through CROP agencies to promote disaster-inclusive policies as one example of joint efforts under the Sendai Framework. The Conferences of the Forum Disability Ministers' Meetings (FDMM), considered coss-sectoral platforms, bring together diverse stakeholders, as exemplified by the endorsement of the PFRPD and the continued involvement of DPOs in articulating and shaping regional priorities (DRF, 2020; PDF, 2025; Sano, 2020).

Sustainable funding and capacity building have empowered DPOs to take the lead in advocacy in the Pacific. One of the results of the grants from DRF/DRAF is that the DPOs in Fiji and PNG could keep their workforce, fortify their organizational frameworks, and partake in advocacy through legislation. Concurrently, the backing of recurring funding has allowed the playing of intricate campaigns, such as making elections in the Solomon Islands more accessible, which has had a positive impact on their efforts. The PDF's strategy signals major funding, empowerment of skills, and the casting of DPOs, particularly women-led groups, into leadership roles; hence, it aligns with Australian aid commitments for CRPD training (DRF, 2020).

Institutional backing is through the impartation of the CRPD knowledge to the political leaders, government focal points, and DPOs as it is stated in the PFRPD Goal 3. The goal is, in particular, the parliamentary sub-committee and model legislation. The DPO Capacity Report of 2013 highlights themes of governance, administration, and leadership and proposes various supportive measures, such as on-the-job training and rural outreach, to boost effectiveness. The USAPI funding through CDC's OT21-2101 is directed at strengthening public health systems through non-profit intermediaries that focus on workforce competencies and infrastructure for inclusive services.

- Grants spread over several years develop the advocacy skills, as DPOs report increased self-assurance in the absence of such assistance.
- Regional bodies like the PFRPD coordinate resource distribution among the FICs.

- Collaborations with UNFPA and UNICEF focus on women with disabilities in terms of SRHR and violence prevention.

The Pacific needs a strong advocacy ecosystem to help more persons with disabilities be included and more resilient. The governments, DPOs, academia, NGOs, and regional bodies have shared information, and through frameworks like the PFRPD, the DPOs have gained significant power in policy-making, data systems, and disaster risk reduction. Such collaborations as DRF/DRAF, PDF, UNFPA, UNICEF, and CDC have supported together DPO leadership, organizational capacity, and participation of women with disabilities. DPOs' geographic collaboration, evidence-based advocacy, and governance involving the disabled will be key factors in the Pacific Region in preventing the disabled from being unseen and in promoting equitable and inclusive development.

8.5 Conclusion

The introduction of disability-inclusive resilience across all fields is not only a question of well-meaning policies; it is a case of different departments taking coordinated actions, tolerance of culture, and making continuous investments. The Asia–Pacific region still has severe disparities in terms of social and economic aspects. On the other hand, it is also giving a positive message evidenced by the rising number of disability-inclusive legislations, better access to ICT, and the rising acceptance of PWDs' input in society. Besides, the megatrends of the day like climate change, digitization, and AI, while gradually transforming societies, still need to be guided by the macroforce of inclusion that denotes these changes in a manner that they come through unquestioned. The Pacific cases illustrate that change occurs when innovators think about access first, when service suppliers collaborate closely with the local communities, and when the local supporters receive the empowerment through specially designed tech solutions that are not required to be on a large scale to be effective. The future of the continent's creativity will depend on the degree of its commitment to accessibility for everyone through the formation of governing bodies that comprehend and practice inclusion, the provision of equitable digital spaces, the elimination of multiple disadvantages, and the allocation of resources for data, research, and technology that reflect the lives of disabled people. The inclusion of every single human being would be the prerequisite for the region to design a future free from nobody being left behind by placing it at the center of every planning, decision-making, and innovation activity.

References

Bharti, N. (2024). *2024 Regional human development report: Making our future: New directions for human development in Asia and the Pacific*. https://www.undp.org/asia-pacific/rhdr2024

Borg, J., Lindström, A., & Larsson, S. (2009). Assistive technology in developing countries: National and international responsibilities to implement the Convention on the Rights of Persons with Disabilities. *The Lancet, 374*(9704), 1863–1865. https://doi.org/10.1016/S0140-6736(09)61872-9

Butcher, N. (2010). *ICT, education, development, and the knowledge society*. GESCI Publisher, 281. https://www.gesci.org/fileadmin/user_upload/4_ICT_in_STEM_Education_Files/ICT__Education__Development__and_the_Knowledge_Society_1__1_.pdf

Chavatzia, T. (2017). *Cracking the code: Girls' and women's education in science, technology, engineering and mathematics (STEM)*. United Nations Educational, Scientific and Cultural Organization. https://doi.org/10.54675/QYHK2407

Colglazier, W. (2015). Sustainable development agenda: 2030. *Science, 349*(6252), 1048–1050. https://doi.org/10.1126/science.aad2333

Commonwealth Scientific and Industrial Research Organisation (CSIRO). (2022). *Seven megatrends that will shape the next 20 years*. https://www.csiro.au/en/news/all/news/2022/july/seven-megatrends-that-will-shape-the-next-20-years

Cooms, S., Muurlink, O., & Leroy-Dyer, S. (2024). Intersectional theory and disadvantage: A tool for decolonisation. *Disability & Society, 39*(2), 453–468. https://doi.org/10.1080/09687599.2022.2071

Cuenca, E. C. (2012, January). The prohibition on discrimination: New content (Art. 14 ECHR and protocol 12). In *Europe of rights: A compendium on the European convention of human rights* (pp. 467–484). Brill Nijhoff. https://doi.org/10.1163/9789004219915_025

Disability Rights Fund, (DRF). (2020). *Evaluation of DRF/DRAF programming in Pacific countries; For Disability Rights Fund*. https://www.disabilityrightsfund.org/wp-content/uploads/DRFDRAF_2017-2019PacificEvalReportSummary_FINAL_May2020.docx

Economic and Social Commission for Asia and the Pacific (ESCAP). (2018). *Building disability inclusive societies in Asia and the Pacific assessing the progress of the Incheon Strategy*. United Nations. https://www.unescap.org/sites/default/files/publications/SDD%20BDIS%20report%20A4%20v14-5-E.pdf

ESCAP, U. (2022a). *A three-decade journey towards inclusion: assessing the state of disability-inclusive development in Asia and the Pacific*. https://www.unescap.org/kp/2022/three-decade-journey-towards-inclusion-assessing-state-disability-inclusive-development

ESCAP, U. (2022b). *Asia-Pacific digital transformation report 2022: Shaping our digital future*. https://www.unescap.org/kp/2022/asia-pacific-digital-transformation-report-2022-shaping-our-digital-future

ESCAP, U. (2022c). *Building forward together: Towards an inclusive and resilient Asia and the Pacific*. https://sdgasiapacific.net/knowledge-products/0000023

ESCAP, U. (2022d). *Framework for disability policies and strategies in Asia and the Pacific*. https://www.unescap.org/kp/2022/framework-disability-policies-and-strategies-asia-and-pacific

ESCAP, U. (2023). *Advancing pacific priorities 2023*. https://www.unescap.org/kp/2023/advancing-pacific-priorities-2023

Global Facility for Disaster Reduction and Recovery (GFDRR). (2017). *Disability inclusion in disaster risk management*. World Bank Group. https://www.gfdrr.org/sites/default/files/publication/GFDRR%20Disability%20inclusion%20in%20DRM%20Report_F.pdf

Gurung, P. (2022). *Indigenous Persons with Disabilities Global Network (IPWDGN)*. Indigenous World.

Heap, B., & Hirmer, S. (2020). Smart villages. *Horizons: Journal of International Relations and Sustainable Development*, (15), 290–305. http://cirsd.org/en/horizons/horizons-winter-2020-issue-no-15/smart-villages

Heaslip, V., & Holley, D. (2023). Ensuring digital inclusion. *Clinics in Integrated Care, 17*, Article 100141. https://doi.org/10.1016/j.intcar.2023.100141

Jevtić, M., Cvetković, V. M., Gačić, J., & Raonić, Z. (2025). *Factors of vulnerability and resilience of persons with disabilities during disasters: Challenges and strategies for inclusive risk reduction.* https://doi.org/10.18485/ijdrm.2025.7.1.6

Joseph, D. D., & Doon, R. A. (2023). *The impact of climate change on vulnerable populations: Social responses to a changing environment* (p. 234). MDPI-Multidisciplinary Digital Publishing Institute. https://doi.org/10.3390/books978-3-0365-5503-4

Kayess, R., Sands, T., & Fisher, K. R. (2014). International power and local action–Implications for the intersectionality of the rights of women with disability. *Australian Journal of Public Administration, 73*(3), 383–396. https://doi.org/10.1111/1467-8500.12092

Labelle, R. (2005). *ICT policy formulation and e-strategy development: A comprehensive guidebook.* Elsevier.

Mendis, K., Thayaparan, M., Kaluarachchi, Y., & Pathirage, C. (2023). Challenges faced by marginalized communities in a post-disaster context: A systematic review of the literature. *Sustainability, 15*(14), 10754. https://doi.org/10.3390/su151410754

Ng, S. B. (2019). *Exploring STEM competences for the 21st century.* 2019, UNESCO International Bureau of Education, IBE/2019/WP/CD/30 REV. https://unesdoc.unesco.org/ark:/48223/pf0000368485

Nguyen-Trung, K., Thuy, T. T. T., Anh, N. P., Cong-Lem, N., Huyen, D. T., Diu, L. T., Giang, N. G., & Simon, M. (2025). Vulnerabilities of people with different types of disabilities in disasters: A rapid evidence review and qualitative research. *Disasters, 49*(3), Article e12686. https://doi.org/10.1111/disa.12686

Organisation for Economic Co-operation and Development (OECD). (2010). *Sickness, disability and work: Breaking the barriers—A synthesis of findings across OECD countries.* https://www.oecd.org/en/publications/2010/11/sickness-disability-and-work-breaking-thebarriers_g1g10adb.html

Pacific Disabilities Forum (PDF). (2007). *Strategic Plan 2007–2011*, http://www.pacificdisability.org/publications.aspx

Pacific Disability Forum, (PDF). (2025). *Preconditions to inclusion issues paper: Introduction to the pacific disability forum's preconditions framework.* https://pacificdisability.org/wp-content/uploads/2024/12/Introduction-to-Precondition-Framework-Issues-Paper.pdf

Park, D. U., Park, H., Kim, Y., Baek, S., Lee, D., Jang, Y., Jung, T., & Rodriguez, F. S. (2021). The role of science, technology and innovation policies in the industrialization of developing countries. *Lessons from East Asian countries, 10.* https://doi.org/10.13140/RG.2.2.20879.87203

Priestley, M., Stickings, M., Loja, E., Grammenos, S., Lawson, A., Waddington, L., & Fridriksdottir, B. (2016). The political participation of disabled people in Europe: Rights, accessibility and activism. *Electoral Studies, 42*, 1–9. https://doi.org/10.1016/j.electstud.2016.01.009

Puamau, P., & Pene, F. (Eds.). (2009). *Inclusive education in the Pacific* (Pacific Education Series, No. 6). Institute of Education, The University of the South Pacific. http://www.usp.ac.fj/index.php?id=ie

Ratliffe, K. T., Rao, K., Skouge, J. R., & Peter, J. (2012). Navigating the currents of change: Technology, inclusion, and access for people with disabilities in the Pacific. *Information Technology for Development, 18*(3), 209–225. https://doi.org/10.1080/02681102.2011.643207

Rechner, L., Harvey, K. E., Lancaster, S., & Horney, J. A. (2023). How COVID-19 impacted people with disabilities: A qualitative study in Delaware. *Public Health in Practice, 6*, Article 100424. https://doi.org/10.1016/j.puhip.2023.100424

Sagala, S., Azhari, D., Nailah, N., Vicri, R. N., Paramitasari, D., Herdiansyah, A. R., & Patricia, C. (2024). Transformative participation in inclusive disaster risk reduction. In *Oxford research encyclopedia of natural hazard science.* https://doi.org/10.1093/acrefore/9780199389407.013.522

Sano, R. (2020). Overview of the ASEAN enabling master plan 2025: Mainstreaming the rights of persons with disabilities and its relevance among ASEAN member states. 現代福祉研究, (20), 21–30. https://doi.org/10.15002/00023394

Schokkenbroek, J. (2004). A new European standard against discrimination: negotiating Protocol no. 12 to the European Convention on Human Rights. In *The development of legal instruments to combat racism in a diverse Europe* (pp. 61–79). Brill Nijhoff. https://doi.org/10.1163/9789004276499_009

Schulze, M. (2010). Understanding the UN Convention on the Rights of Persons with Disabilities. *Advocate, 1*, 1–4. http://www.handicap-international.fr/fileadmin/documents/publications/HICRPDManual.pdf

Sharma, U., Jitoko, F., Macanawai, S. S., & Forlin, C. (2018). How do we measure implementation of inclusive education in the Pacific Islands? A process for developing and validating disability-inclusive indicators. *International Journal of Disability, Development and Education, 65*(6), 614–630. https://doi.org/10.1080/1034912X.2018.1430751

Spurway, K., & Griffiths, T. (2016). Disability-inclusive disaster risk reduction: vulnerability and resilience discourses, policies and practices. In *Disability in the global south: The critical handbook* (pp. 469–482). Springer International Publishing. https://doi.org/10.1007/978-3-319-42488-0_30

Thongkuay, S. (2024). Strengthen voices of women and girls with disabilities in Asia and Pacific. In *World congress on rehabilitation 2024*. https://doi.org/10.54878/6bsp8z62

Tjondronegoro, D., Kendall, E., Hunter, S., & Mayer-Besting, E. (2025). *Reimagining disability: Is inclusion a macroforce that will shape Asia and the Pacific for the next 20 years?*. https://doi.org/10.25904/1912/5785

Twigg, J., Kett, M., & Lovell, E. (2018). *Disability inclusion and disaster risk reduction* (ODI Brief. Note, 1–12). https://doi.org/10.13140/RG.2.2.20114.91846

Umar, A. (2018). Smart collaborating hubs and a smart global village-an alternative perspective on smart cities. In *2018 IEEE Technology and Engineering Management Conference (TEMSCON)* (pp. 1–6). IEEE. https://doi.org/10.1109/TEMSCON.2018.8488404

UNDRR. (2023). *Thematic Report on Disability Inclusion in Disaster Risk Reduction in the Pacific.* https://www.undrr.org/media/89314/download?startDownload=20251206

United Nations Economic and Social Commission for Asia and the Pacific (UNESCAP). (2002). *Biwako Millennium Framework, United Nations publication, Sales No. E.01.XVII.15.* http://www.unescap.org/esid/psis/disability/bmf/bmf.html

United Nations, Economic and Social Commission for Asia and the Pacific (ESCAP). (2023). *Leveraging digital innovation for inclusive and sustainable development in Asia and the Pacific. United Nations ESCAP, Social Development Division (SDD), Bangkok.* https://www.unescap.org/knowledge-productsseries/social-development-working-papers

Women, U. N. (2022). *Progress on the sustainable development goals: The gender snapshot 2022.* https://www.unwomen.org/en/digital-library/publications/2022/09/progress-on-the-sustainable-development-goals-the-gender-snapshot-2022

Zian, C., & Oguzhan, C. L. (2023). *Leveraging digital innovation for inclusive and sustainable development in Asia and the Pacific.* https://www.unescap.org/knowledge-productsseries/social-development-working-papers

Appendix A
Glossary of Terms

Ableism: Discrimination or social prejudice against persons with disabilities based on assumptions of able-bodied norms.

Accessibility: The degree to which products, environments, programs, and services are usable by all people, including those with disabilities.

Adaptive Capacity: The ability of individuals or communities to adjust to hazards, moderate damage, and take advantage of opportunities.

Assistive Device: Equipment or tools that enhance the functional abilities of persons with disabilities (e.g., mobility aids, hearing devices).

Assistive Technology (AT): Digital or mechanical technologies designed to support daily functioning and participation.

Baseline Data: Initial measurements used to compare future changes in vulnerability or resilience.

Biopsychosocial Model: A model of disability integrating biological, psychological, and social factors.

Climate Mitigation: Efforts to reduce greenhouse gas emissions or enhance carbon sinks.

Collective Resilience: Community-level ability to withstand and recover from shocks.

Community-Based Disaster Risk Management (CBDRM): Localized planning and action to reduce disaster

S. M. Thomas and R. Veerabathiran, *Disability, Disaster, and Resilience in Oceania*, SpringerBriefs in Modern Perspectives on Disability Research,
https://doi.org/10.1007/978-981-95-8535-9

impacts.

CRPD (Convention on the Rights of Persons with Disabilities): A UN treaty protecting the rights and dignity of persons with disabilities.

Cultural Competence: The ability to respectfully engage with diverse artistic practices, beliefs, and values.

Differential Vulnerability: The unequal exposure to risk among population groups due to socio-economic and structural factors.

Digital Divide: Inequalities in access to digital technologies, tools, and the internet.

Disability-Inclusive Disaster Risk Reduction (DIDRR): Ensuring that DRR planning and action fully include persons with disabilities.

Disaster Governance: Structures and processes that guide disaster planning, response, and recovery.

Disaster Risk: The potential loss of life, livelihood, or assets due to hazard exposure and vulnerability.

Disaggregated Data: Data broken down by categories such as disability, gender, age, and location.

Early Warning System (EWS): Systems that detect and communicate threats to enable timely and effective response.

Ecosystem-Based Adaptation (EbA): Using biodiversity and ecosystem services to reduce climate risks.

Framework for Resilient Development in the Pacific (FRDP): A regional guide integrating climate adaptation and DRR.

Functional Limitations: Restrictions in physical, cognitive, sensory, or psychological abilities.

Gender Mainstreaming: Integrating gender perspectives into all levels of policy and planning.

Geospatial Mapping: Use of geographic data to analyze spatial patterns and risk.

Governance Systems: Political and institutional structures guiding decision-making.

Human Rights-Based Approach (HRBA): Integrating human rights principles into policy, programs, and governance.

ICT (Information and Communication Technologies): Digital tools enabling communication, data flow, and information access.

Identity Politics: Social movements or claims-making based on shared identities (e.g., disability rights).

Inclusion: Ensuring equal participation, dignity, and opportunity for all individuals.

Indigenous Knowledge Systems (IKS): Traditional ecological and cultural knowledge rooted in ancestral practices.

Intersectionality: The overlapping systems of inequality (e.g., gender, disability, ethnicity) shaping experiences.

Kinship Networks: Extended family systems are central to Pacific social organization and resilience.

Knowledge Co-Production: Collaborative creation of knowledge by researchers, communities, and stakeholders.

Marginalization: Processes that exclude individuals or groups from full participation.

Monitoring and Evaluation (M&E): Methods used to assess the effectiveness of programs and policies.

Multi-Hazard Approach: Planning for multiple, interacting disaster risks simultaneously.

Ocean Governance: Management of marine resources and ecosystems in the Pacific.

Organizations of Persons with Disabilities (OPDs): Rights-based disability groups advocating for inclusion.

Participatory Planning: Involving affected communities directly in decision-making.

Policy Implementation Gap: The gap between written policies and actual practice.

Protection Cluster: Humanitarian structure coordinating protection efforts for vulnerable populations.

Rapid-Onset Disaster: Sudden hazards such as cyclones or earthquakes.

Relational Worldview: A Pacific cultural understanding that relationships, land, and community shape identity.

Resilience: The ability to anticipate, absorb, adapt, and recover from shocks.

Sendai Framework: Global DRR framework emphasizing inclusive governance and risk reduction.

Social Determinants of Health: Conditions shaping health outcomes, including inequality and access to services.

Social Exclusion: Processes that prevent individuals from participating fully in society.

Stakeholder Engagement: Collaboration among government, community, private sector, and civil society.

Te Ao Māori: Māori worldview centered on interconnectedness, spirituality, and kinship with land and sea.

Traditional Ecological Knowledge (TEK): Indigenous knowledge guiding environmental stewardship.

Twinning Partnerships: Collaborations between disability organizations across regions to build capacity.

Universal Design: Designing environments and tools accessible to all people regardless of ability.

Urban Vulnerability: Increased exposure to hazards due to dense populations and infrastructural strain.

Vulnerability Assessment: Analysis of the conditions that increase susceptibility to harm.

Vulnerability Spectrum: Recognition that vulnerability varies by social, cultural, and economic context.

Women with Disabilities (WWDs): A group facing intersectional discrimination based on gender and disability.

World Risk Index: A global comparative metric of disaster risk combining exposure, susceptibility, and coping capacity.

Youth Engagement: Inclusion of young people in planning and resilience-building.

Appendix B
Figures and Tables

1. **Figure 1:** Accessibility Challenges Across the Disaster Cycle
   This figure illustrates the key accessibility challenges experienced by persons with disabilities throughout the disaster management cycle. During the preparedness phase, barriers arise from inaccessible early warning systems and the absence of disability-disaggregated data. In the response phase, non-inclusive evacuation procedures and a shortage of trained responders create significant risks. The mitigation phase is marked by limited participation of persons with disabilities in planning processes and the lack of universal-design infrastructure. In the recovery phase, reconstruction efforts often overlook accessibility needs, and the loss of assistive devices frequently remains unaddressed.
2. **Figure 2:** Comparison of existing disability-related barriers and the improvements required to achieve inclusive disaster management systems
   This figure compares the existing barriers that persons with disabilities face during disaster situations with the improvements needed to close the disability inclusion gap. The left panel illustrates current challenges, including inaccessible risk information, environmental hazards, stress, and exclusion in information access, non-inclusive healthcare facilities, and the loss or inaccessibility of assistive devices. The right panel highlights necessary improvements such as accessible urban infrastructure, inclusive communication technologies, community participation in preparedness, disability-inclusive healthcare systems, and the integration of universal design principles in rebuilding and planning. This comparison emphasizes the need for systemic changes to create resilient and inclusive disaster management frameworks.
3. **Figure 3:** Interaction of climate hazards, structural inequities, and disability barriers creating heightened risks and adverse outcomes for PwD
   This figure illustrates how climate hazards, structural factors, and disability-specific barriers intersect to intensify disaster-related risks for persons with disabilities. The left panel highlights climate threats, and the central panel shows

S. M. Thomas and R. Veerabathiran, *Disability, Disaster, and Resilience in Oceania*, SpringerBriefs in Modern Perspectives on Disability Research,
https://doi.org/10.1007/978-981-95-8535-9

structural challenges, including poverty, inadequate infrastructure, limited healthcare access, and unsafe housing conditions. The right panel depicts disability-related barriers. Together, these overlapping factors contribute to compounded outcomes, including higher injury rates, difficulty evacuating, exclusion from disaster planning, loss of assistive devices, and increased mortality.

4. **Figure 4:** Key rights-based interventions needed to strengthen gender- and disability-inclusive disaster risk reduction. This figure highlights a set of rights-based interventions essential for advancing inclusive disaster risk reduction for women with disabilities.
   These interventions include implementing international commitments such as CRPD, CEDAW, and the Sendai Framework; improving access to sexual and reproductive health (SRH) and WASH services; and ensuring the participation of women with disabilities in decision-making processes related to disaster planning and response. Additional priorities involve expanding livelihood opportunities, developing inclusive education systems, strengthening land rights, and recognizing intersectionality between gender and disability. Collectively, these measures contribute to a more equitable and rights-based approach to disaster resilience.
5. **Figure 5:** Layers of environmental, social, historical, and cultural factors shaping disability-related vulnerability in disaster contexts.
   This figure illustrates how multiple interconnected factors contribute to the heightened vulnerability of persons with disabilities during disasters. Environmental hazards such as cyclones, droughts, and volcanic eruptions interact with socioeconomic challenges like limited resources and geographic isolation. These vulnerabilities are further compounded by colonial legacies, including land dispossession and institutional structures that shape present-day inequities. Additionally, cultural beliefs, oral traditions, and spiritual practices influence community responses and perceptions of disability. Together, these layered factors create a complex risk environment that disproportionately affects individuals with disabilities.
6. **Figure 6:** Overview of the P-CEP Framework outlining capabilities, person-centered principles, and iterative steps for personalized emergency planning for persons with disabilities. This figure presents the Personalized-Capability Emergency Planning (P-CEP) Framework, which supports individuals with disabilities in developing tailored emergency preparedness plans.
   The framework integrates three key components: (A) a capability assessment that considers communication, mobility, assistive devices, health needs, support networks, transportation options, and environmental factors; (B) person-centered planning principles emphasizing collaboration, self-determination, shared decision-making, respect for lived experience, strength-based approaches, and inclusive communication; and (C) four iterative steps involving identifying needs and strengths, creating a personalized emergency plan, connecting with support networks and services, and regularly practicing,

reviewing, and updating the plan. Together, these components help ensure inclusive, effective, and sustainable preparedness for persons with disabilities.

7. **Table 1:** Colonial Mapping Practices and Their Effects on Indigenous Geographies
8. **Table 2:** Policy and Implementation Gaps in Pacific Island Disaster Governance
9. **Table 3:** Dimensions of Double Vulnerability Among Women with Disabilities
10. **Table 4:** International and Regional Frameworks for the disabled persons in the Pacific

Appendix C
Methodological and Cultural Considerations

1. Respect for Indigenous Epistemologies
   Pacific knowledge systems—including oral histories, communal decision-making, and ecological observation—are treated not as supplementary but as primary sources of knowledge. The interpretation of lived experience is therefore contextualized within cultural worldviews such as: *vanua* (Fiji): land, people, and spirit as interconnected; *fa'a Samoa* (Samoa): collective identity and mutual responsibility; *kastom* (Melanesia): customary governance and relational authority.
2. Emphasis on Lived Experience
   Testimonies of persons with disabilities serve as foundational data points. Their perspectives were interpreted through narrative analysis, disability rights frameworks, and community-based experience mapping.
3. Reflexivity and Positionality
   The research acknowledges power dynamics between researchers, communities, and disabled participants. Care was taken to avoid imposing Western disability constructs, prioritize local definitions of ability, identity, and wellbeing, and recognise community ownership of knowledge.
4. Ethical and Cultural Protocols
   Data interpretation follows cultural protocols such as obtaining consent within relational settings, respecting customary authority structures, and ensuring confidentiality in small island contexts.
5. Limitations
   Accessibility to remote island regions, diverse cultural interpretations of disability, and limited disaggregated data pose challenges to generating region-wide generalizations.

S. M. Thomas and R. Veerabathiran, *Disability, Disaster, and Resilience in Oceania*, SpringerBriefs in Modern Perspectives on Disability Research,
https://doi.org/10.1007/978-981-95-8535-9

Appendix D
Relevant Frameworks and Policies

1. International Frameworks

   - Convention on the Rights of Persons with Disabilities (CRPD): Mandates non-discrimination, accessibility, participation, and equal protection.
   - The Sendai Framework for Disaster Risk Reduction (2015–2030) highlights disability-inclusive early warning, preparedness, and governance.
   - The Sustainable Development Goals (SDGs) include disability-specific indicators: SDG 10 (Reduced Inequalities) and SDG 11 (Sustainable Cities).
   - UNFCCC & Paris Agreement: Addresses climate impacts disproportionately affecting vulnerable groups.

2. Regional Frameworks

   - Pacific Framework for the Rights of Persons with Disabilities (PFRPD): Sets standards for accessibility, protection, leadership, and community inclusion.
   - Framework for Resilient Development in the Pacific (FRDP): Integrates climate adaptation, DRR, and sustainable development.
   - Incheon Strategy to "Make the Right Real" for Persons with Disabilities in Asia and the Pacific: A 10-goal roadmap for regional disability inclusion.

3. National Policy Examples (PICs)

   - Disability Acts and National Policy Frameworks (Fiji, Samoa, Vanuatu, PNG)
   - National Adaptation Plans (NAPs)
   - Social Protection and Inclusive Education Acts
   - National Disaster Management Plans (NDMPs)

S. M. Thomas and R. Veerabathiran, *Disability, Disaster, and Resilience in Oceania*, SpringerBriefs in Modern Perspectives on Disability Research,
https://doi.org/10.1007/978-981-95-8535-9

Appendix E
Statistical Insights

1. Disability Prevalence in the Pacific
   - An estimated 15–18% of Pacific Islanders live with a disability.
   - Some regions (especially remote and outer islands) report prevalence rates exceeding 20% due to limited healthcare access and underreporting.
2. Disaster Exposure
   - PICs rank among the top 20 most disaster-prone countries globally.
   - Cyclones, droughts, floods, volcanic eruptions, and sea-level rise disproportionately affect rural and disabled populations.
3. Accessibility and Services
   - Less than 30% of persons with disabilities have access to formal assistive devices.
   - Fewer than 25% report having access to fully accessible evacuation shelters.
4. Gender-Disability Intersections
   - Women with disabilities face two to three times higher risk of experiencing gender-based violence during and after disasters.
   - Employment rates for disabled women remain below 20% in many PICs.
5. Digital Inclusion
   - Internet penetration varies widely—from 15% in some remote islands to 80% in urban centers.
   - Only one in five disabled Pacific Islanders regularly accesses digital platforms for emergency information.

S. M. Thomas and R. Veerabathiran, *Disability, Disaster, and Resilience in Oceania*, SpringerBriefs in Modern Perspectives on Disability Research,
https://doi.org/10.1007/978-981-95-8535-9

6. Policy and Participation

   - While most PICs have ratified the CRPD, fewer than 40% have fully implemented disability-inclusive DRR strategies.
   - Disabled persons' organizations (DPOs) are involved in national policy consultation processes in only half of the countries surveyed.

MIX
Papier aus verantwortungsvollen Quellen
Paper from responsible sources
FSC® C105338

If you have any concerns about our products,
you can contact us on
ProductSafety@springernature.com

In case Publisher is established outside the EU,
the EU authorized representative is:
Springer Nature Customer Service Center GmbH
Europaplatz 3, 69115 Heidelberg, Germany

Printed by Libri Plureos GmbH
in Hamburg, Germany